NOT FOR GREENS

not *for* greens

He who sups with the Devil should have a long spoon

Ian Plimer

With a foreword by Patrick Moore, co-founder of Greenpeace

Connor Court Publishing Pty Ltd

PO Box 224W
Ballarat VIC 3350
sales@connorcourt.com
www.connorcourt.com

ISBN: 9781925138191 (pbk.)

Cover design by Ian James

Printed in Australia

CONTENTS

About the author

PROFESSOR IAN PLIMER is Australia's best-known geologist. He is Emeritus Professor of Earth Sciences at the University of Melbourne, where he was Professor and Head of Earth Sciences (1991-2005) after serving at the University of Newcastle (1985-1991) as Professor and Head of Geology. He was Professor of Mining Geology at The University of Adelaide (2006-2012) and in 1991 was also German Research Foundation research professor of ore deposits at the Ludwig Maximilians Universität, München (Germany). He was on the staff of the University of New England, the University of New South Wales and Macquarie University. He has published more than 120 scientific papers on geology and was one of the trinity of editors for the five-volume *Encyclopedia of Geology*. This is his eighth book written for the general public, the best known of which are *Telling Lies for God* (Random House), *Milos-Geologic History* (Koan), *A Short History of Planet Earth* (ABC Books), *Heaven and Earth* (Connor Court) and *How to Get Expelled from School* (Connor Court).

He won the Leopold von Buch Plakette (German Geological Society), the Clarke Medal (Royal Society of NSW), the Sir Willis Connolly Medal (Australasian Institute of Mining and Metallurgy). He is a Fellow of the Australian Academy of Technological Sciences and Engineering and an Honorary Fellow of the Geological Society of London. In 1995, he was Australian Humanist of the Year and later was awarded the Centenary Medal. He was Managing Editor of *Mineralium Deposita*, president of the SGA, president of IAGOD, president of the Australian Geoscience Council and sat on the Earth Sciences Committee of the Australian Research Council for many years. He won the Eureka Prize for the promotion of science, the Eureka Prize for *A Short History of Planet Earth* and the Michael Daley Prize (now a Eureka Prize) for science broadcasting. He was an advisor to governments and corporations and was a regular broadcaster.

Professor Plimer spent much of his life in the rough and tumble of the zinc-lead-silver mining town of Broken Hill where an interdisciplinary scientific knowledge intertwined with a healthy dose of scepticism and pragmatism are necessary. His time in the outback has introduced him to those who can immediately see the weaknesses of an argument. He is Patron of Lifeline Broken Hill and the Broken Hill Geocentre. He worked for North Broken Hill Ltd and was a director of CBH Resources Ltd. In his post-university career he is proudly a director of a number of listed (Silver City Minerals Ltd, Niuminco Group Ltd, Sun Resources NL, Lakes Oil NL and Kefi Minerals plc) and unlisted companies (Roy Hill Holdings Pty Ltd, Hope Downs Iron Ore Pty Ltd and Queensland Coal Pty Ltd).

A new Broken Hill mineral, plimerite $ZnFe_4(PO_4)_3(OH)_5$, was named in recognition of his contribution to Broken Hill geology. Ironically, plimerite is green and soft. It fractures unevenly, is brittle and insoluble in alcohol. A ground-hunting rainforest spider *Austrotengella plimeri* from the Tweed Range (NSW) has been named in his honour because of his "provocative contributions to issues of climate change". The author would like to think that *Austrotengella plimeri* is poisonous.

Acknowledgements

Various people have been ear-bashed for years on this subject and some concepts in this book I aired at various professional conferences for criticism and specialist feedback. A draft was read a number of times by Andrew Drummond who checked calculations and put a huge amount of effort into structure and context. Sir Jim R. Wallaby (aka Barry Williams) twice used his experienced editorial eye to suggest changes, additions and corrections. Mining engineer Richard Fitzpatrick, metallurgist Bob Greenelsh, geologists Peter Kitto, Fiona McKenzie and John Nethery all made useful comments in their areas of expertise. Viv Forbes and Mischa Popoff made constructive comments on cropping and farming. Michael Darby, Phil Sawyer and my pedant colleague Kyle Wightman all had a number of comments and suggestions on an early draft. Alexandra Nicol provided additional information for Chapter 2 and made constructive comments. My normal critics Bob Besley and my wife Maja added their two bob's worth of constructive comments on an early draft. My former student Rhonda O'Sullivan gleefully settled scores by finding errors, correcting English kinda like and providing constructive comments on two versions. I am grateful to these folk for being critics, editors and readers under time pressures and I have no idea whether they agree or not with the ideas expressed in this book. I am grateful to Patrick Moore who wrote the Foreword under great time pressures.

As per usual, Anthony Cappello of Connor Court had to deal with me, a task that deserves a knighthood. Previous books were written to a background of ABC Classic FM but this radio station has now devolved into a facile yabber-fest with the playing of motel elevator music. *Not For Greens* was written with background web-streamed delightful music from Radio Swiss Classic.

FOREWORD

Congratulations to Ian for producing a book that provides a different, and in my opinion extremely rational, look at the agenda of the green movement today. In many respects they have become a combination of extreme political ideology and religious fundamentalism rolled into one. There is no better example of this than the fervent belief in human-caused catastrophic climate change.

It is the duty of every scientist to remain vocally skeptical of theories that have not been "proven" by the scientific method. Except with regard to "climate change", of course, where it is apparently the duty of every scientist to adhere to the strictest dogma, because "the science is settled", so keep your trap shut.

Thus Ian Plimer is a heretic for daring to question the current belief that humans are responsible for most of the minor warming that has occurred during the past 150 years, which, if it continues will certainly be the downfall of civilisation and will cause ecological destruction of untold proportions. And there will be no silver lining in that cloud.

Such is the state of affairs that sceptics are named "deniers", as in Holocaust deniers, even though the Holocaust is an historical fact whereas climate change is about the entirely different task of predicting the future. Predicting the future with computer models, no less, as if they are a crystal ball in the hands of a fortune-teller. They seem to forget that the crystal ball is a legend, a mythological thing, and not an actual future-predicting-device. As Yogi Berra famously opined, "Predictions are difficult, especially about the future."

The basis of the scientific method is the observation of natural phenomena, directly with our eyes or with instruments we have devised to detect things that are beyond our natural senses. When we believe we are observing a new thing or phenomenon we must report it in the scientific literature in such a way that other competent scientists can attempt to replicate it. If they do we are on the path towards a "proof", that what we reported is indeed valid.

The certainty among many scientists that humans are the main cause of climate change, including global warming, is not based on the replication of observable events. It is based on just two things, the theoretical effect of human-caused greenhouse gas emissions, predominantly carbon dioxide (CO_2), and the predictions of computer models using those theoretical calculations. There is no scientific "proof" at all.

This is why the International Panel on Climate Change uses the word "likely" in their statement about human-caused warming. True, they say "extremely likely" but putting a strong adjective in front of "likely" does not change the reality that they are expressing an opinion, not a fact. They even use the term "expert judgment" to describe their opinion. Upon close analysis it is obvious to a sceptical mind that the entire issue of climate change is more about belief than it is about science.

Consider a few actual facts:

- The concentration of CO_2 in the global atmosphere is lower today, even including human emissions, than it has been during most of the existence of life on Earth.
- The global climate has been much warmer than it is today during most of the existence of life on Earth. Today we are in an interglacial period of the Pleistocene Ice Age that began 2.5 million years ago and has not yet ended.
- There was an Ice Age 450 million years ago when CO_2 was about 10 times higher than it is today.
- Humans evolved in the tropics near the equator. We are a tropical species and can only survive in colder climates due to fire, clothing, and shelter.

- CO_2 is the most important food for all life on earth. All green plants use CO_2 to produce the sugars that provide energy for their growth and our growth. Without CO_2 in the atmosphere carbon-based life could never have evolved.
- The optimum CO_2 level for most plants is about 1600 parts per million, four times higher than the level today. This is why greenhouse growers purposely inject the CO_2-rich exhaust from their gas and wood-fired heaters into the greenhouse, resulting in a 40-80 percent increase in growth.
- If human emissions of CO_2 do end up causing significant warming (which is not certain), it may be possible to grow food crops in northern Canada and Russia, vast areas that are now too cold for agriculture.
- Whether increased CO_2 levels cause significant warming or not, the increased CO_2 levels themselves will result in considerable increases in the growth rate of plants, including our food crops and forests.
- There has been no further global warming for nearly 18 years, during which time about 25 percent of all the CO_2 ever emitted by humans has been added to the atmosphere. How long will it remain flat, and will it next go up or back down? Now we are out of the realm of facts and back into the game of predictions.

I am particularly pleased with Ian's use of the stainless steel teaspoon as an example of a simple tool that is the product of many technologically advanced procedures. It reminds me of the fact that my former organisation, Greenpeace, is today opposed to all mining. If you ask them for the name of any mine that is operating in an environmental acceptable standard, you will draw a blank. They have become so cornered by their own extremism that they must deny their daily use of cell phones, computers, bicycles, rapid transit, and yes, the simple teaspoon. Mining is one of the fundamental bases of human civilisation.

All our metals, much of our energy, and nearly all our built infrastructure are the products of mining materials from the earth. When it comes to our survival the only major exceptions are our food and fiber crops, the cultivation of which also requires many tools that are the product of mining. Yet Greenpeace and their allies see no use for it.

As Ian so thoughtfully exposes, we humans would be much better off, and so would the environment, if we ignored the doomsayers and continued to work at balancing environmental, social, and economic priorities, the aim of sustainable development. Every day more than seven billion people wake up on this planet with real needs for food, energy, and materials. Sustainability is partly about continuing to provide for those needs, and at the same time learning to do so in ways that reduce damage to the environment. It is the only sensible path forward.

Dr Patrick Moore,
Co-founder, Greenpeace,
May 6, 2014

1

INTRODUCTION

This book is deliberately offensive

"It's not easy being green" carolled Kermit the Frog in his chart-topping single some years back and, while I hesitate to disagree with such a distinguished salientia, I must say that until recently in the body politic of Australia it cannot have been too hard. It certainly didn't require much in the way of deep thought, knowledge, science or self-doubt. And the uncritical media scientifically-illiterate lapped it up. But the auguries seem to suggest that the bubble is about to burst, and not before time.

It is hardly a startling claim that most people prefer to live in a pleasant environment, though what constitutes pleasant is as diverse as the range of people around the world. But most reading this would probably include a certain "greenness" in their definition. Trees, shrubs, grass and flowers are generally perceived as pleasant. I would agree.

But "green" itself is not necessarily a good or bad thing. Green vegetables are good for you but green meat should be avoided at all costs. There are many toxic green plants. Green emeralds are very pretty and not toxic. There are many other pretty green minerals that are toxic, and copper (II) acetoarsenite (Paris Green), while pretty, is also highly toxic. Some green snakes are pretty but can kill you.

And so we get to the concept of green as applied to politics. What started as a laudable movement to prevent the despoilation of certain areas of natural beauty has morphed into an authoritarian, anti-progress, anti-democratic, anti-human monster. The age of innocence is over. We can't take everything at face value even though it might be on a web site or

Wikipedia. There is no substitute for knowledge that has been validated and is in accord with other validated knowledge. And if you don't have this knowledge, then a little common sense will get you a long way.

On the surface, the green aims appear to be to revert our species to living in some previous idyllic age, which in fact never existed. To do this greens seem determined to eschew, and to force everyone else to comply, every benefit that human science, enterprise, ingenuity and innovation have bestowed on us and which has contributed overwhelmingly to the pleasantness most of us, in the West, enjoy today. We see ignorance hiding behind slogans.

The purpose of this book is to demonstrate just how much of our wonderful pleasant environment has been a consequence of human adaptability, ingenuity and enterprise, and just how disastrous the acceptance of wholesale unquestioned "green" utopianism would be for our species and our planet. We are witnessing the tip of the iceberg of green utopianism touting "alternative" sources of electricity from wind and solar energy regardless of the effects on the environment, communities and the bottom line.

This book is about perspective, common sense and the obvious. I use a single implement that we all use in everyday life, a stainless steel teaspoon, to demonstrate that our quality of life depends on humans being able to think, adapt and implement in the face of threats to our wonderful world. The real threat to our world comes from fanatical political green ideology and not from environmental degradation.

In the world of mindless marketing, we hear that if it's green, it must be good. In this book I argue that this is not the case, despite the vacuous rent-a-celebrities attaching their names to causes they don't understand. The greens want to save everything and also the world. And so do I: from greens, green-initiated unemployment, green policies that create high-energy costs and green hubris. The green political parties claim that they are parties for people, progress and protest. In this book, I show that it is the exact opposite.

If something is supported by the greens, then it is assumed that it must be for sound reasons and these reasons are backed by the latest science. In this book, I show that this mantra is a fallacy.

So, is this book deliberately offensive? If you are an ideological green, then yes. However, I am comforted by the fact that greens don't read books because they already know all that is to be known. If you have your feet on the ground, have the thirst for knowledge, are rational and have your mind anchored to reason, then no. If you are a genuine environmentalist, then no. And, if you are looking for answers in a morass of conflicting ideas, propaganda, politics, ideology and information, then this book will provide you with some of the answers. Mind you, the best way to get answers is to use that increasingly rare commodity on Earth: common sense.

This book is deliberately offensive to greens. They prove over and over again that they underpin their ideology by a lack of knowledge, hypocrisy and dreamy impracticable solutions. Greens are committed to problems, not solutions. They object to any solution and angrily reject any evidence that a problem may not be as bad as they purport to fear but, in reality, for which they hope. Many green schemes meant to save the planet have proven to be environmentally devastating and I show a few examples in this book. This book is directed at greens who try to take away our freedoms, destroy the thousands of years of accumulated knowledge, challenge the foundations of our culture, create high costs of living for the average person and keep populations in poverty.

This book is directed at greens who have no understanding of how planet Earth operates, who erroneously claim that emission of traces of a trace gas by humans is the driver of climate change and accordingly want to change society by totalitarian means. I argue that greens have no skin in the game yet want to control society without having faced the electorate. If you find this offensive, then my answer is: good. It's true and it's about time that unelected squeaky prophets and profiteers were challenged robustly.

The normal greens' answer to criticism is *ad hominem* attacks. These attacks are in the absence of cogent alternative arguments based on validated evidence to support their causes. Past experience shows that the greens' response to criticism is to plagiarise and repeat the rantings of others who have also not read scientific books of criticism. My green critics surf the web, find www.I-am-an-idiot.com, quote *ad nauseum* and yet never check the original source or critically evaluate what they read or write. Why bother when there is something to quote that agrees with your ideology.

For example, Wikipedia has some wonderful howlers because of a lack of checking the original source and omissions that may be contrary to editors' ideology. The green echo chamber of groupthink blogs and web sites is characterised by vulgar, profane, violent and sexist abuse in the absence of reasoned argument. If Twitter is your medium, then there can be no reasoned argument in 140 characters, only a demonstration of failed intellectual and character tests. So, dear reader, I am left with you and I am sure that you would describe yourself as an environmentalist. As I do.

Why on Earth would anyone write about a stainless steel teaspoon? Or a knife, plate, pot, pan or any other item we use in everyday life? A stainless steel teaspoon is very much a symbol of the modern world and there were thousands of years of progress applied to produce a cheap, simple and hygienic teaspoon. The stainless steel teaspoon is a symbol of how we have escaped from crushing poverty, rampant disease and life-threatening pollution. We take all this for granted.

There are good reasons why we use cutlery and don't eat with our hands. It was science, engineering, risk and capitalism that gave us the stainless steel teaspoon. I await the time when greens lead by example and eat with their hands, do not use reticulated water and electricity and live a sustainable life in the wilderness. Until then, we have no logical reason to take them seriously.

Maybe individual greens are inconsequential people with few

achievements who think that doing something for the environment will make them feel good about themselves? As I show, greens create poverty and destroy wealth. Without wealth, there are no funds for the environment. Societies that accommodate their village idiots are robust; a society that listens to them is suffering from "progressive" politics.

In order to undertake the mineral exploration needed to find the iron, chromium, nickel and possibly manganese, molybdenum and tin needed for your stainless steel teaspoon, physical and financial risks are taken and the latest high-level scientific and engineering knowledge and skills are used. Nature is fickle so exploration is a combination of interpretative sciences and is normally unsuccessful. Exploration uses energy. The commodities required to make a stainless steel teaspoon are not all found in any one single country and, in order to use the end product in our everyday life, international trade, transport, risk and finance are required. If the commodities for your stainless steel teaspoon are found by exploration, they need to be shown to be of adequate size and metal content to profitably mine.

In some developing countries, the political risk is such that negotiation may be at the end of a Kalashnikov and only concluded by bribes, influence or a bigger gun. In Western countries political risks are from ignorance, uncertain policy, lack of leadership and green pressure and come at the end of a bureaucrat's pen. Risks can be partially lowered by increasing knowledge, experience and skills.

Mining uses energy, technical knowledge and constantly evolving skills. Once the commodities for your stainless steel teaspoon are discovered and mined then concentration and beneficiation are needed. Energy is consumed. Commodities need to be transported, smelted and formed into a stainless steel alloy. The process of smelting has evolved over thousands of years from trial and error to a high precision process that uses a massive amount of energy. After stainless steel has been made, your spoon needs to be fabricated and transported so that it can be purchased locally at a price that will provide a profit for the local store owner.

Before you actually use your stainless steel teaspoon to eat, every stage of the process to create it involves risk, international finance and trade. Every stage must be profitable. Every stage involves a collective human effort by individuals who are not connected and who are not even aware that their effort is part of the process of your eating food. Every stage involves humans trying to earn a living and trying to make a better world for the next generation. You cannot use a stainless steel teaspoon (or any other item in the modern world) unless you embrace exploration, mining, smelting, international trade, international finance, risk-taking capitalism, integration of the latest applied science and engineering and low-cost efficient energy.

Greens do not invest in serious employment-producing businesses, do not start significant businesses, do not take risks with their easily-earned cash and have little idea of how a capitalist economy works. Very few green leaders have worked in private enterprise. Let me give green readers a hint. If you live off welfare or a government job, then for every $1 you are given, someone has to take all the risk, expend energy through working, pay tax and earn $5. The same person who tries to earn the $5 to keep greens alive is being attacked by green ideology and quite often has to retrench workers because of green activism. Many greens also attack profit. Without profit there could be no taxation. Without taxation there could be no green schemes.

In effect, the greens object to the world being a better place than it was in the past. They use modern communication systems to promote their ideology yet that communication is only possible because of cheap energy and metals. Some might say the greens are hypocritical. Others may say they are ignorant of the basics. I argue that they are both and furthermore that they are a malevolent unelected group attempting to deconstruct healthy societies that have taken thousands of years to evolve.

We are all environmentalists and do not want to pollute the air, water, soils and life. We want the next generations to have better lives than ours.

We all want a clean planet and a high standard of living. We want the freedoms that have taken generations of struggle to achieve. However, we are not all greens. Some of us very strongly object to an organised noisy negative minority dominating a positive poorly-organised majority that just wants to get on with a peaceful life. Some of us strongly object to this noisy minority who want to strip us of our freedoms in order to impose their unproven and impracticable ideology. Some of us object strongly to a noisy minority wanting to direct an economy wherein their contribution is, at best, modest. I am yet to see greens that have skin in the economic game.

The greens want to create a mythical idyllic lost world by destroying thousands of years of innovation, science, engineering and culture. This mythical *Nirvana* never existed. The past was pretty bleak with disease, starvation, early death, few freedoms and poverty. During periods of global cooling, people died like flies. In periods of global warming, societies thrived. A romantic view of that mythical long lost wonderful past is to be ignorant of history. We live in an unprecedented time of peace since the fall of the Roman Empire when food and energy are abundant.

Throughout this book I show that the greens are not at all interested in the environment but are interested in retrogressive politics, control of people and an abuse of freedoms. I argue in this book that their behaviour shows that they are ignorant of the basic processes that give us the modern world yet they are quite happy to be ignorant and hypocritical beneficiaries of that world. I also argue that generating electricity by so-called "renewable" energy sources such as wind, solar or biomass cannot provide enough energy to create the humble stainless steel teaspoon that we all use today.

A stainless steel teaspoon contains a large amount of embedded energy, mostly derived from coal. The chemical process of reduction to make metals for your stainless steel teaspoon derives from coal. The greens assert that the production of energy by burning fossil fuels leads to carbon dioxide emissions into the atmosphere. (Right). The greens then assert that the emitted carbon dioxide (a greenhouse gas) drives global warming. (Wrong). The argument is extended to speculate that this global warming will be sudden, irreversible and catastrophic. (Wrong). They then assert that this emitted carbon dioxide leads to extreme weather events. (Wrong).

The greens have kindly taken it upon themselves to save us from global warming and consequently want emissions of carbon dioxide reduced or stopped. This is a proxy for stopping industries that give us light, heat, cooked food, refrigeration, air conditioning, transport, holidays and consumer items such as your humble stainless steel teaspoon. This is the action of authoritarian socialists commonly known as watermelons (green on the outside, red inside). This is not an environmental view, it is a political view contrary to that of most of the community who are aspirational and just want to get on with life with the least amount of bother.

Although I have an interest in climate, as one who has spent decades teaching science I have a far greater interest in the intellectual climate of society. Green arguments about global warming commonly use logical fallacies. One argument is that the majority of scientists support the concept of human-induced global warming. Besides having no data to support this assertion, it is a logical fallacy. Just because many people believe something to be true does not make it true. Aristotle showed this 2,350 years ago.

Another logical fallacy is resorting to authority. It is argued that many authorities such as the Intergovernmental Panel on Climate Change (IPCC), the Royal Society, etc, claim that there is human-induced global warming. However, there are just as many authorities that argue that there is no evidence to support this view yet such dissenting views are belittled and scarcely mentioned by the scare-mongering media and the greens. This leaves us with the impression that everyone agrees with their apocalyptic scenario.

It is often touted that 97% of scientists consider global warming to be real and caused by human-induced emissions of carbon dioxide. This derives from a survey sent to 10,257 people, of which the 3,146 respondents were further whittled down to 77 self-selected climatologists of whom 75 were judged to agree that human-induced warming was taking place. What is not mentioned is that in the US there were more than 30,000 scientists who signed a document claiming that human-induced global warming was nonsense. I am not playing a numbers game here, as I am only too well aware of Aristotle's logical fallacies. My point is that there is selective reporting of information and surveys.

By contrast, the American Physical Society (APS) which with more than 50,000 members is probably the largest society of physicists in the world, is formally examining competing views on the physical science basis of global warming. In the light of the failure of IPCC climate change models to predict reality, the APS has appointed a six-person panel including three independent scientists of a sceptical nature.

Even the IPCC now agrees with the UK Met. Office that temperature rise has ceased. As carbon dioxide is still increasing and temperature is not, therefore this gas could not be driving global warming. This scenario has been validated from geology and ice core measurements of past similar events. The temperature has not increased as the 90 IPCC climate models predicted and the satellite and surface temperature measurements are not in accord with 95% of these models.

The choice is simple. Do we accept theoretical models of future temperature or do we accept actual temperature measurements? The IPCC has solved this little problem. They have increased the hype about forthcoming disasters despite earlier stating that the predicted disaster wasn't actually happening. Yawn. Some 2,350 years ago, Aristotle would have argued that it is not an authority that determines the truth of a matter, it is evidence. Nothing changes.

Well, what about human emissions of carbon dioxide causing global climate change? Myth or reality?

Carbon dioxide and climate change

Let me emphatically state that I am not sceptical that climate changes. It always has changed and always will. Notwithstanding, greens paint me as a "climate denier", whatever this means. Carbon dioxide is a greenhouse gas and has a very slight effect on the atmosphere at current or elevated concentrations. However, I am sceptical that human emissions of carbon dioxide drive climate change because the empirical evidence from the history of planet Earth shows that natural climate changes have been rapid, large and unrelated to carbon dioxide, let alone human activity. Although humans may have a slight effect on the Earth's atmosphere, the Earth's atmosphere does not drive climate change. It is the medium through which climate is propagated and human effects are swamped by the enormous natural changes on Earth. But first, a bit of Earth history.

If a carbon dioxide molecule is released into the atmosphere, it only hangs around for about five to seven years before it is sequestered into the oceans or by life. The oceans contain a lot of heat, far more than the atmosphere. The top 3.2 metres of the oceans contain the same amount of heat as the whole atmosphere. The key to climate change is the oceans, not the atmosphere.

Many greens seem to think that once a carbon dioxide molecule is released into the atmosphere, there it stays and it is not recycled as part of the larger integrated carbon and water cycles. This would mean that carbon dioxide is an inert atmospheric gas like helium, neon, argon, krypton or xenon. But then again, no one is claiming that greens have a basic understanding of science.

Earth history

Our planet is a wet warm volcanic planet. For most of time, planet Earth was warmer and wetter than now. We are actually in an ice age that commenced 34 million years ago within which there are cyclical glacials and interglacials. We are currently in an interglacial.

Our planet formed 4,500 million years ago by the condensation and recycling of stardust associated with the formation of an exceptionally stable star in a good galactic address. Some say this happened on a Thursday while others claim that it happened on a Monday. The latter group are called "Thursday deniers". Early in the history of the Solar System, Earth was bombarded by massive asteroids. Every time a primitive sea formed by condensation of volcanic steam, it was vaporised by impacting. This was the limitation to the formation of life on Earth. As soon as the surface and atmosphere of the Earth had sufficiently cooled and asteroid bombardment decreased, rainwater accumulated and life formed. The evolution of life is inextricably linked to a very weird molecule. It is water. Without water, there would be no volcanoes, no recycling of crustal rocks, no oceans, no climate change and no life.

Life

First life on Earth formed at least 3,800 million years ago and, by 3,500 million years ago, this bacterial life had colonised the planet. These bacterial colonies are still with us. Bacteria were the first life on Earth, they are still the dominant life on Earth, they have survived all the natural catastrophes on Earth and they make up the largest biomass on Earth. It may come as a shock to greens but neither whales nor trees are the greatest biomass on Earth. I have never heard green activists wanting to save bacteria.

It was not until Earth was middle-aged that the first great ice ages occurred. For less than 20% of time, Earth has had ice sheets. The first and second of six great ice ages occurred when the continents were clustered around the equator, the Earth was covered in ice and ice sheets occurred at sea level at the equator. At these times the Earth was either a snowball or a slush ball.

Around 2,500 million years ago a number of irreversible events took place. The continents grew thicker, nutrients washed from retreating ice sheets to fertilise the oceans, bacterial life diversified and the atmosphere started to accumulate oxygen excreted from bacteria. At that time, the atmosphere contained some 30% carbon dioxide and oxygen caused a mass extinction of prokaryotic bacterial life. Limey rocks and carbon-bearing sediments started to become abundant. This sedimentation started the 2,500 million year drawdown of carbon dioxide from the atmosphere. This drawdown is still taking place and, in the history of the planet, we are living in times when the atmospheric carbon dioxide content is very low.

A giant supercontinent started to fragment 830 million years ago releasing massive quantities of water vapour, carbon dioxide and methane into the atmosphere. Ice sheets again formed at sea level at the equator 800 million years ago, there were sea level rises and falls of up to 1,500 metres and the oceans were fertilised by nutrients from retreating ice sheets. Again, there was a diversification of life and, because the oxygen content of the atmosphere had risen slightly and the oceans had more nutrients, multicellular marine animals formed 583 million years ago (Ediacaran fauna).

The Ediacarans grazed on sea floor algal mats but 542 million years ago, multicellular animals developed shells, skeletons and protective coatings because there were enough nutrients in the oceans required for development of the necessary muscle functions. This explosion of life created a massive drawdown of carbon dioxide from the atmosphere. It took 20 million years for most of the major animal groups to evolve. These armoured animals quickly destroyed the unarmoured Ediacarans.

Mass extinctions

Some 50 million years later, land plants appeared. Land plants have existed for only 10% of Earth history. The first of the major mass extinctions of complex life took place 440 to 420 million years ago associated with asteroids impacting. Life quickly recovered and thrived to fill the vacated ecological niches. There was a minor ice age after which life thrived. Another major mass extinction occurred 360 million years ago. An asteroid impact in Sweden, massive global volcanism and a mass extinction all occurred at the same time. Life quickly recovered and plants again thrived. A result of the post-impact thriving was that the atmospheric carbon dioxide content decreased, the methane content increased and the oxygen emitted by abundant vegetation increased to the point where it was common for the Earth's atmosphere to spontaneously ignite.

Continental drift continued and the supercontinent Gondwana (now South Africa, India, Australia, South America and Antarctica) drifted over the South Pole. The southern continents were covered by ice sheets that waxed and waned. During the warmer interglacials thick accumulations of waterlogged vegetation formed peat bogs. This peat was later compressed to form the coals that are mined today to provide the energy to make your stainless steel teaspoon. While the southern continents were enjoying this ice age, the northern continents were equatorial and had drifting red sand dunes, shallow water coral reefs and shelf sediments rather similar to the modern Persian Gulf.

This suddenly changed 251 million years ago with a mass extinction of 96% of all complex species on Earth. Complex life on Earth nearly became extinct. There is great speculation about what would have happened if all life had become extinct. Would the identical evolutionary path have occurred or would life on Earth be significantly different? The jury is still out. This mass extinction probably resulted from monstrous volumes of sulphurous gases released from huge Siberian volcanoes. These gases killed land-based vegetation and animals and temporarily acidified the top few metres of the ocean killing floating marine organisms and shallow marine life.

Earth recovered to being a normal warm wet planet, only to be interrupted by another impact-induced mass extinction of life some 217 million years ago. A giant supercontinent started to fragment and the Atlantic Ocean formed. This fragmentation caused major degassing of water vapour, carbon dioxide and methane from deep in the Earth. Plant life thrived with the extra atmospheric carbon dioxide and, as a result, animal life thrived. Earth again recovered from a mass extinction, life continued to diversify and the continents continued to drift.

Another mass extinction of life occurred 65 million years ago. This is the one we all know about. Dinosaurs and many other creatures became extinct. One theory proposes that an asteroid hit Mexico causing ejected fragments, dust and sulphurous gases to be thrown into the atmosphere. A global dust layer is preserved in rocks that formed at this time. The planet would have been dark and the atmosphere full of sulphurous gases. Vegetation, herbivores and floating marine organisms died.

Another theory is that the massive volcanoes that gave us the Deccan Traps of India ejected huge volumes of sulphurous gases into the atmosphere. Again, life would have died from the sulphurous fumes in the atmosphere. Nothing is settled in science. Life recovered from this mass extinction, as it always has.

India had been happily drifting across the Indian Ocean and finally collided with Asia some 50 million years ago. This collision resulted in the pushing up of the Tibetan Plateau. Local weather and climate were changed. New mountains were stripped of soils and new soils formed in the lowlands extracting carbon dioxide from the atmosphere. These new soils were buried as sediment. Meanwhile, the bare rock in the alpine areas created monsoonal updrafts that dragged in warm moist air from the Indian Ocean to produce monsoonal rains. This still happens today.

Today's ice age

South America had the good sense to pull away from Antarctica 37 million years ago. Consequently a circum-polar current isolated Antarctica from warm water and an ice sheet formed some 34 million years ago. This ice sheet waxed and waned, as it does now. This is what ice sheets do. They expand and contract and just because an ice sheet may be contracting in our lifetime does not necessarily mean that we humans have anything to do with a normal geological process. There are more than 20 active volcanoes underneath the Antarctic ice sheet. These emit much heat but are not considered by climate activists in their sea ice and ice sheet models. Why not?

An overall cooling trend commenced and the slight changes in the Earth's orbit created cycles of relative cooling and warming. Global cooling drove human evolution and migration over the last five million years. The Earth's elliptical orbit varies cyclically over a 100,000-year period, its axis changes over a 41,000-year period and it wobbles like a top over a 21,000-year cycle. Each of these processes varies the Earth's distance from the Sun resulting in a cyclical pattern of warming and cooling. I await green-driven legislation to stop these cycles of orbitally-driven global climate change.

Closure at Panama of the connection between the Atlantic and Pacific Oceans 2.67 million years ago prompted further global cooling. Coincidentally, a supernoval eruption caused the Earth to be bombarded by cosmic radiation and thick clouds formed. This exacerbated the cooling caused by the increased orbital distance of the Earth from the Sun. As a result, the Greenland ice sheet formed. The weight of ice in both Greenland and Antarctica pushed down the centre of both these land masses to form basins with raised rims. Consequently, ice must first flow uphill before flowing over the rims into glaciers draining towards the sea. The outward directed pressure caused by the central accumulation is not caused by temperature. The waxing and waning of ice sheets and glaciers are extraordinarily complicated by various factors and air temperature is only a minor contributor.

In the past, climate fluctuated between warm and cold periods every 41,000 years. About one million years ago a 100,000-year cycle commenced. This involved, on average, a 90,000-year cold glacial period followed by a 10,000-year warm interglacial period. We are currently about 10,500 years into a warm interglacial period that peaked around

6,000 years ago. In the period from 12,000 to 6,000 years ago, sea level rose about 130 metres at an average of two centimetres *per annum*. Global temperature has decreased about 2°C and sea level has fallen nearly two metres since the 6,000-year interglacial peak.

The future climate

The current ice age has not finished and we have no idea when the Earth will return to being a normal warm wet greenhouse planet. Although interglacials usually last about 10,000 years, we do not know when the current one will end even though it has been in progress for more than 10,000 years. However, we cannot escape the fact that the current interglacial will end and we will enjoy another 90,000 years of glaciation. The past gives us a clue about a glaciated planet and there is no reason to suggest that the next glaciation will be any different.

Previous glaciations had kilometre-thick ice sheets that covered Canada, northern USA, most of Europe north of the Alps, most of Russia and elevated areas in both hemispheres. Much of the Andes, New Zealand and Tasmania were covered by ice. Upland areas, even in the tropics, had glaciers. In areas with no ice sheets, strong cold dry winds shifted sand and devegetation occurred. Dunes in Australia, North Africa, the Middle East and North America again moved and great wind-deposited loess deposits covered Mongolia, China and northern USA.

In Australia, the Great Barrier Reef, the poster child of the greens, disappeared during glacial events more than 60 times over the last three million years. It reappeared after every one of these events. The Great Barrier Reef first formed about 50 million years ago and has survived hundreds of coolings and warmings and massive rain events that deposit sediment on the Reef. The sea level fall and lower temperature during glacial events kills higher latitude coral reefs and they continue to thrive at lower latitudes. The geological record shows that coral reefs love it warm, especially when there is more carbon dioxide in the atmosphere.

During glacial events, tropical vegetation is reduced from rainforest to grasslands with copses of trees, somewhat similar to the modern dry tropics inland from the Great Barrier Reef. There was no Amazonian rainforest during the last glaciation.

The Sun drives the Earth's climate. There is a demonstrated relationship between solar activity and climate and solar physicists note that the Sun's activity is declining. When inevitable solar- or orbitally-driven cooling happens again, we will have the normal conflicts, depopulation and devegetation that occur every time the planet is cooler. A colder planet means a hungrier planet. We will then long for global warming.

This chronicle of our planet is a sensual evocative story underpinned by empirical evidence. No computer models have been able to replicate it 20 years in advance let alone centuries in advance. The atmospheric system alone is so complicated that meteorologists cannot accurately predict the weather a month in advance. No climate model has ever been able to replicate the past. With new evidence, this chronicle of the history of our planet is continuously refined. Many very large forces have interacted to produce this chronicle. Some forces were random, others were cyclical and others were probably irreversible. Much is still unknown. One fact is certain: it is not the trace gas carbon dioxide that has driven past climate changes, most of which have been greater and more rapid than any changes measured in modern times. Why should carbon dioxide now drive climate change? Is it because we now have greens?

What we do know is that the planet is constantly changing and such changes need to be factored into changes that the green ideology claims are due to recent human emissions of carbon dioxide. For example, over the last 250 years rainfall over England and Wales has increased by 5%. Might this slight rise be due to the warming since the Maunder Minimum ceased 300 years ago? The greens need to show that this variability is not natural and is due to human activities. They have not.

Ever since the planet formed, there has been climate change. If we humans wanted to change climate on Earth, we would have to stop bacteria doing what bacteria do, change ocean currents, manage the drift of continents, change the Earth's orbit, control the variability of the Sun and control supernoval explosions. And the greens actually believe they can stop climate change. This is egocentric and clearly delusional. In former times, such people would have been locked up.

Climate change

Climate cycles

Planet Earth is dynamic. Climate change, sea level change, extinctions, changing ocean currents and waxing and waning of ice sheets are normal. In the past, climate cycles were of galactic (143 million years), orbital (100,000, ~41,000 and ~21,000 years), solar (1,500, 210, 87, 22 and 11 years), oceanic decadal (~30 years) and lunar tidal (~18.6 years) origin. Oceans have decadal oscillations (e.g. Pacific Decadal Oscillation [PDO]) and non-cyclical major events (e.g. *El Niño-La Niña*) when the surface temperature of ocean water changes. Asteroid impacting and volcanicity are the major non-cyclical climate-changing events. There is no evidence to show that the past was different from the present.

If the Earth's climate did not constantly change, then I would be really worried.

Oceans

The heat capacity of water establishes that it is not the temperature of the atmosphere that heats the ocean surface waters. It is the temperature of the ocean surface that heats the atmosphere. This can easily be shown at home. If one has a hot bath, the heat from the water heats the whole bathroom. If one has a bathroom heater, the warm air does not heat the water.

The 3,000 ARGO buoys deployed a decade or so ago are showing that the surface temperature of the ocean is decreasing yet the carbon dioxide content of the atmosphere is increasing. Again, there is a disconnect between atmospheric carbon dioxide and temperature.

At the time of writing, there have been 17 continuous years when global atmospheric temperature has not increased despite the increasing human emissions of carbon dioxide, especially by China. Again, further evidence that carbon dioxide does not drive global warming. If it plays a part, then it is minor and masked by much larger natural processes.

Galactic cycles

Galactic climate cycles derive from increased bombardment of the Solar System with cosmic rays. These cosmic rays induce the formation of low-level clouds that reflect heat. The Earth then cools. The six major ice ages that planet Earth has enjoyed coincided with its location in the Sagittarius-Carina (twice), Perseus, Norma, Scutum and Orion Arms when there was increased cosmic radiation. Wobbles in the Earth's orbit produce cycles of warm and cold resulting from changes in the distance between the Earth and the Sun.

The Sun has a number of regular cycles and outbursts of energy. These influence climate because they result in changes in the solar magnetic field that, in turn, protects the Earth from cosmic ray bombardment. It may appear heretical to those advocating human-induced climate change, but that great ball of energy in the sky that we call the Sun actually drives surface energy systems and life on Earth.

Supervolcanoes

Climate changes have sporadically occurred as a result of supervolcanoes, supernoval eruptions, tectonism and possibly impacts. Large volcanic eruptions at tropical latitudes (e.g. Tambora, Indonesia, 1815; Krakatoa, Indonesia, 1883) have ejected aerosols into the stratosphere that reflect light and heat to produce cooler stormy weather that lasts for years. The Laki (Iceland) eruption in 1783 was a small eruption yet it resulted in the bitterly cold winter of 1783-1784, freezing of harbours and rivers (such as the Mississippi River at New Orleans), ice in the Gulf of Mexico, summer frosts and biting cold winds. The resulting famine in Iceland killed 24% of its population.

The 10 April 1815 eruption of Tambora resulted in crop destruction in the US, multiple summer frosts and snow over much of the US agricultural areas, halving of the summer growing season and famine. In Europe, there were grain shortages as crop yields decreased by 75%, grain prices rose by up to 300% in June 1817 and many in Europe stayed alive by eating rats, cats, dogs, horses, grass and straw. The same will happen with another big volcanic eruption and solar- or orbitally-driven cooling. Grain exports from the world's wheat belts would greatly reduce and there could be mass starvation in places that import large quantities of grain such as North Africa, the Middle East, Yemen, Afghanistan and Iran. Past periods of cooling led to famine, war and changes in the global power structure. Will the next cooling do the same?

Mount Pinatubo eruption on 15 June 1991 was another small eruption that resulted in the cooling of the planet in 1992 by 0.5°C. Although this does not seem much, in the northern Canadian wheat belt, crops did not reach maturity before winter and only hay was harvested. Is this a window into future global cooling?

Terrestrial supervolcanoes (e.g. Yellowstone, USA; Taupo, NZ; Kamchatka, Russia; Toba, Indonesia) have ejected thousands of cubic kilometres of aerosols into the atmosphere and these can trigger or accelerate global cooling. Planet Earth started to cool 116,000 years ago at the beginning of the most recent glaciation. During this cooling, Toba (Indonesia) erupted 74,000 years ago. It filled the atmosphere with about 3,000 cubic kilometres of aerosols and accelerated the rate of global cooling. Humans very nearly became extinct. The same filling of the atmosphere with aerosols also occurs after an asteroid impact.

The Sun

When thinking about climate, there are two fundamental factors to consider.

The first factor is the atmosphere, which is the medium through which the climate manifests itself. On Earth it consists of (in round figures) 78% nitrogen, 21% oxygen; 1% argon, 0.04% carbon dioxide and traces of other gases. There are varying amounts of water vapour. Most planets and many moons have primitive atmospheres, and thus climates, but they are all different. Furthermore, Earth didn't always have the atmosphere and climate it has now. The Earth's atmosphere, climate, life and the continents have evolved over time and continue to evolve.

The second fundamental factor is the energy input into the atmosphere that causes the climate to be what it is. In the case of Earth (and the rest of the Solar System), energy input is from that giant fusion reactor we call the Sun. The Sun is the nuclear engine that, overwhelmingly, drives everything on the surface of the Earth, including the climate, per medium of the atmosphere. If greens are anti-nuclear, then they should object to solar power.

While the Sun is a remarkably stable star, as stars go, it is by no means static. It, too, has an atmosphere, which, apart from being very hot, is also very turbulent. It is quite capable of throwing out immense clouds of hot, ionised gases many millions of kilometres into space, sometimes with drastic effects on both the Earth's atmosphere and on spacecraft travelling outside the lower atmosphere and the Earth's protective magnetic shield.

Activist arguments about climate change tend to concentrate very largely on the medium (atmosphere), while the inputs of the engine (Sun) are usually taken as a given. The atmosphere holds very little energy, the greatest amount of surface energy derived from the Sun resides in the oceans. In fact, we cannot hope to begin to understand the climate unless we take both these fundamental factors, and not just carbon dioxide, into account.

To hear the rather alarming and definitely simplistic propositions being bandied about in the political space, we would think that all the answers about our changing climate are related to our carbon dioxide in the atmosphere. The argument is presented that the effects of the Sun are so well known that we need consider them no further. You do not need to be a scientist to see that such arguments are facile, to say the least. As Marshall McLuhan could have said: "The medium is only part of the message."

In truth, there are huge unknowns about our atmosphere and even more about our Sun, not to mention the interaction between them. That's OK, that's what science is all about, finding out about what we don't know and, possibly more importantly, validating what we do know. Science is a very humbling pursuit.

Deniers

There has never been a public debate about human-induced climate change. Only dogma. Science is full of different interpretations of similar observations and, while it sometimes leads to heated and protracted arguments (scientists being as human as the next *Homo sapiens*), it seldom leads to one side trying to equate their opponents with all the basest characteristics of the human species. Yet this is precisely what happens in the climate change non-debate. Question even one minor factor in the "official" story and you are likely to be accused of all sorts of political chicanery and moral turpitude. I am yet to find a scientist or read a paper that claims climate is not changing. Hence, to label someone as a climate change "denier" demonstrates that the accuser believes that without human activity, climate would not change. This is ignorance.

This doesn't happen in, for example, the arcane mysteries of the equally complex field of quantum mechanics where there are at least a dozen interpretations, each with its champions. If these practitioners attack each other with cries of "denier", "traitor", "tool of big business" or similar abhorrent epithets, they certainly do not appear as leading stories in the print or electronic news media.

But more critically, disputes between Copenhagenists and Baseanites, no matter how heated, do not usually spill over into the political arena leading to shrill, hasty and irrational policies, with deleterious effects on both the economy and life in general.

A pity the same cannot be said of the climate change non-debate.

Planetary degassing and carbon dioxide

Since the formation of planet Earth, there has been degassing from the mantle and deep crust of water vapour, carbon dioxide, methane, sulphur-bearing gases, nitrogen and many other gases. This has been caused by molten rocks solidifying at depth (plutonism), by molten rocks reaching the surface (volcanism) and by rocks being cooked up at high pressure and high temperature (metamorphism).

Volcanic venting of carbon dioxide

Degassing also occurs on other planets, as does climate change. Degassing of carbon dioxide occurs before, during and after plutonism and volcanic eruptions from gas vents, hot springs and craters. Some 1,800 terrestrial volcanoes are known and only about two dozen are accurately monitored. Submarine degassing occurs from at least 3.47 million off axis submarine basaltic volcanoes and from volcanic activity along the entire 64,000-kilometre length of mid ocean ridges.

At mid ocean ridges, the oceanic crust of the Earth is pulled apart by tectonic forces allowing gases from deep in the mantle to leak to the surface. Each year some 10,000 cubic kilometres of seawater circulates through new hot mid ocean ridge basalt as a coolant. This heats the oceans. This also keeps the oceans alkaline partly because of chemical reactions between hot basaltic rocks and seawater. The ocean floor also has many cross cutting or transverse fractures where there are no volcanoes. These fractures vent large quantities of carbon dioxide leaking from the mantle.

Basalt is the volcanic rock on the sea floor. Experimental studies show that carbon dioxide is highly soluble in basalt. Basalts collected from the sea floor, although containing carbon dioxide, show evidence they have vented most of the carbon dioxide before eruption. Carbon dioxide is far less soluble in the melts erupted from terrestrial volcanoes hence they exhale less carbon dioxide. To use measured data on emissions of carbon dioxide from a few terrestrial volcanoes to show that the Earth's volcanic emissions are low, as has been claimed, is deliberately misleading. The Earth and other planets derived their atmospheres from degassing. Where do the greens think atmospheric gases might come from?

In 1999, a slow spreading mid-ocean ridge (Gakkel Ridge, Arctic Ocean) experienced an explosive submarine basaltic eruption. For basalt to explode at such a great water depth, at least 13.5 weight per cent of the molten rock must have been carbon dioxide. The erupted volcanic rocks were cooled by circulating seawater and the Arctic Ocean warmed for a short time. This warming was coincidental with a lunar tidal node that pushed warmer surface North Atlantic Ocean water into the Arctic. There is not one permanent deep submarine temperature and volcanic exhalation measuring station hence the emissions of heat and carbon dioxide from submarine basaltic volcanism can only be deduced. In some places, liquid carbon dioxide has been found on the ocean floor and carbon dioxide gas vents are very common.

Furthermore, submarine carbon dioxide released from fractures, gas vents, hot springs and submarine basalt eruptions does not bubble up to the surface as it dissolves in cool high-pressure bottom waters and is degassed to the atmosphere thousands of years later when these waters rise in zones of upwelling.

Carbon dioxide venting from depth

Degassing also occurs from rising bodies of molten rock that freeze kilometres from the surface (i.e. plutons). During ascent, they undergo constant degassing of steam, carbon dioxide, methane and sulphurous gases as do wet sediments and limey rocks. Plutonic and volcanic rocks mainly occur where areas of the Earth's crust are pushed together, such as the current collision of Africa with Europe. In this setting, mountains also form (e.g. European Alps) and, as rocks are converted by heat and pressure to metamorphic rocks, gases such as steam, carbon dioxide and methane are released. These commonly form the spa waters typical of alpine areas.

There are a whole group of rocks from the mantle that explode to the surface as gas-driven volcanoes. The gas is carbon dioxide. Kimberlites, the host to diamonds, are blasted from at least 150 kilometres depth to the surface by expanding carbon dioxide from the mantle. Gas expansion sometimes has been so energetic that some of these kimberlite volcanoes might even erupt at temperatures below 100°C because expansion requires energy which is taken from the kimberlite and carbon dioxide. Carbonate minerals formed by chemical interaction between rock and carbon dioxide are very common in kimberlites.

Another mantle rock is carbonatite. Most of the world's rare earth, niobium and strontium minerals are mined from carbonatites. Carbonatites were molten carbonate rocks with a very large amount of carbon dioxide dissolved in the melt. They form conical masses and may flow out on the surface. The last observed carbonatite lava eruption was at Ol Doinyo Lengai (Tanzania) in 1990. Compared with other lava eruptions, the lava temperature is low (400 to 500°C) and carbon dioxide is vented during eruption. The mantle constantly leaks carbon dioxide to the Earth's atmosphere, mainly in the deep oceans.

The geological evidence shows that the mantle contains abundant carbon dioxide and this has been vented to the atmosphere since the beginning of time and, contrary to the mantra of some green "scientists", is still a major contributor to the atmosphere. The same mantle degassing takes place on other planets in our Solar System.

Human emissions of carbon dioxide in the atmosphere

According to greens, carbon dioxide is a miracle gas. From the green perspective, a very slight increase of carbon dioxide in the atmosphere can have great effects on climate and weather but only if it is from sinful human emissions and not from natural emissions. This miracle gas can simultaneously cause extreme heat and extreme cold, flooding rain and endless drought, increased snow and a lack of snow, increased wind and a lack of wind, and increased hurricanes and a lack of hurricanes. The real miracle of carbon dioxide is that, without this trace gas, there would be no life on Earth. This seems to have escaped the greens' attention. Maybe it hasn't. Maybe they are being disingenuous with the scientifically illiterate media.

Human emissions of carbon dioxide need to be placed in perspective. If annual total emissions of carbon dioxide comprise 33 molecules, only one is from human emissions and the rest from natural processes. This one molecule of human-derived carbon dioxide is mixed with 85,000 other molecules of other gases in air. Carbon dioxide is a trace gas in the atmosphere and humans add traces of a trace gas to an existing trace gas in the atmosphere. If human emissions of carbon dioxide drive climate change, then it has to be demonstrated that this one molecule in 85,000 drives climate change and that the 32 molecules of carbon dioxide derived from natural processes do not. No wonder they use easily manipulated computer models because empirical evidence gives an answer contrary to their ideology.

It has yet to be shown that human emissions of carbon dioxide drive climate change. In fact, there is only evidence to the contrary.

Laws of physics

The first 100 parts per million (ppm) of carbon dioxide in the atmosphere have the greatest effect as a greenhouse gas and the effect decreases exponentially beyond that concentration. A molecule of carbon dioxide has a residence time in the atmosphere of five to seven years before it is sequestered into shells, rocks, plants and ocean water. The atmosphere has 400 ppm carbon dioxide, so a doubling or quadrupling of human emissions of carbon dioxide will have very little effect on temperature unless atmospheric carbon dioxide residence times can be ideologically persuaded to change to two orders of magnitude higher.

The Beer-Lambert Law shows that if the current atmospheric carbon dioxide content is doubled, global atmospheric temperature will increase by 0.2°C and if the carbon dioxide content is quadrupled, temperature will rise by a further 0.1°C. In fact, if all of the world's fossil fuels were burned, the atmospheric carbon dioxide content would not even double. We would have to burn every limey rock on Earth to significantly increase the atmospheric carbon dioxide content. It has yet to be shown that the atmospheric increase in carbon dioxide is due to natural degassing or human emissions of carbon dioxide. In the geological past, the atmosphere contained 30% carbon dioxide compared to the 0.04% today. Yet there were no tipping points, catastrophes, extreme weather or runaway global warming. In fact, each of the six major ice ages was initiated at a time when the atmospheric carbon dioxide content was far higher than at present.

As atmospheric carbon dioxide increases, more and more dissolves in the ocean waters (Henry's Law). Ice core measurements show that during the past six interglacials, temperature rises some 650 to 1,600 years before the atmospheric carbon dioxide content rises, hence a rise in temperature drives atmospheric carbon dioxide content, not the inverse. Some greens seem to think that if we humans emit a single molecule of carbon dioxide, it stays in the atmosphere forever. Nothing could be further from the truth. There are natural cycles that constantly recycle carbon, water and other materials on our planet.

However, something is seriously wrong with the greens' ideology about the effects of carbon dioxide. The atmospheric carbon dioxide content has risen by just over 10% over the last 25 years but the RS satellite global lower troposphere temperature anomaly record shows that warming over that period is statistically indistinguishable from zero. Will the greens please explain? Furthermore, since 1950, there has been an average global temperature rise of 0.6°C and we have been constantly told that polar bears are facing extinction because of this warming. Can the greens please explain why there are now at least five times as many polar bears than there were in 1950? If the greens claim that it has been due to less hunting, then they are admitting that they have been telling lies.

Recent events

In more modern times, planet Earth enjoyed the Roman and Medieval Warmings when there were 600 and 400 years respectively of times far warmer than now. These warmings were separated by the cold Dark Ages when glaciers advanced, crops failed, famine was rife, the weakened population succumbed to the plague and there was massive depopulation. During the Roman and Medieval Warmings, the rate of post-glacial sea level rise did not accelerate, glaciers retreated, there was no sudden emission of carbon dioxide into the atmosphere from industrialisation and it was so warm that the Romans were able to grow crops at latitudes where such cultivation now would not be possible. In the Medieval Warming, the Vikings grew barley, wheat, sheep and cattle on Greenland in areas where farming today would be impossible. Viking graves on Greenland were deep and dug in soil showing that there was no permafrost at that time, as there is today. These warmings were global and unrelated to human emissions of carbon dioxide.

The Little Ice Age started in 1300 AD, the coldest periods were those of no sunspot activity (i.e. the solar magnetic field was small and allowed the ingress of more cosmic radiation). Temperatures fluctuated wildly, there were short warm periods interspersed with long cold periods, glaciers advanced in cooler times and retreated slightly in warmer times, crops failed, people starved, the plague struck Europe in 1347 AD and there was massive depopulation.

The Central England Temperature record, the longest continuous set of temperature measurements began in 1659. The temperature rise from an average of 7.8°C in 1696 to 10°C in 1732 is a huge rise of 2.2°C over 36 years. In modern times, the greatest temperature increase was 0.7°C over the last 100 years. There was no major carbon dioxide-emitting industry in the late 17th and early 18th Centuries. After this warming, there was cooling followed by a slow temperature rise and the 10°C peak in average temperature in 1732 was not reached again until 1947. Can the greens please explain how this temperature rise was both three times as large and three times as fast as in the 20th century? It certainly was not due to human emissions of carbon dioxide. No matter where we look, the green ideology is not underpinned by evidence.

The coldest period of the Little Ice Age, the Maunder Minimum, was 300 years ago. Since then, planet Earth has been warming and there is no answer to the key question. Which part of the post-Maunder Minimum warming is natural and which part is of human origin? Until this question can be answered quantitatively, then there is no measured evidence for human-induced global warming, only computer speculations. There have been more than two decades of computer speculations and this has been enough time to show that the climate has not followed the climate computer models and cannot be predicted, even over a short period of time.

It has been shown many times that these computer model projections are gross over exaggerations. For example, the IPCC regularly resorts to fraud. In the IPCC's Working Group II Summary for Policymakers, they state: "Climate change over the 21st century is projected to increase displacement of people (medium evidence, high agreement)". This statement drove a media frenzy about a forthcoming apocalypse. The Final Draft IPCC WGII AR5 report actually states:

> It is difficult to establish a causal relationship between environmental degradation and migration … Many authors argue that migration will increase during times of environmental stress … and will lead to an increase in abandonment of settlements … Another body of literature argues that migration rates are no higher under conditions of environmental or climate stress.

The Summary claims that climate change "will increase risks of violent conflicts" and the body of the Final Draft says the opposite: "Research does not conclude that there is a strong positive relationship between warming and armed conflict." Journalists are either far too lazy to critically read the Final Draft or just write any exaggerated story that supports their own green dogma. Journalists should be sceptical of everything, hold politicians to account, not do the bidding for green politicians and not close down debate.

The UK Met. Office consistently gets its seasonal forecasts hopelessly wrong. The barbecue summer forecast of 2009 was a washout, the October 2010 forecast that December would be warmer than average preceded the coldest December ever and the March 2012 prediction of a dry April was followed by the wettest April on record. The November 2013 prediction was for a drier than average winter. It was not. It was very wet. The same computer procedures that predict we are going to fry and die in 2100 are used to make seasonal forecasts.

Since thermometer measurements were recorded, temperature decreases (1880-1910, 1940-1977, 1998-present) and increases (1860-1880, 1910-1940, 1977-1998) show that there is no correlation of temperature with increasing atmospheric carbon dioxide. The planet has not been warming rapidly, changes measured today are well within natural variability and even in recent times of industrialisation, there has been both cooling and warming. With no observable correlation between global warming and atmospheric carbon dioxide on geological, ice core and historical time scales, there can be no causation.

Pick your own climate

Claims that one particular year is the warmest on record are misleading and deceptive, especially as these claims are made uncritically and use "corrected" data. Faith trumps facts. Journalists are supposed to be sceptical about all sorts of claims on all matters. However, claims that the planet is warming or doomed are repeated because, on climate matters, most journalists have abandoned scepticism, lack scientific analytical skills and have become apocalypse advocates on behalf of the greens.

To argue that temperature and sea level are increasing depends on when measurements first started and depends on what is measured. If you want to show that the planet is warming, pick the period 1977-1998. If you want to show that the planet is warming and that this warming is natural, pick the last 300 years. If you want to show that the planet's surface temperature is not changing, pick the last 17 years. If you want to show that the planet's surface is cooling, pick the last 6,000 years. If you want to show that climate naturally varies, then look at the last 1, 10, 100 or 1,000 million years as I do.

Even this 17-year figure is arguable, depending upon the measurement method. With satellite measurement techniques, there has been no significant warming for the last 17 (UAH) or 24 years (RSS). For surface temperature measurements, there has been no significant warming for the last 17 years (Hadsst2), 18 years (Hadcrut4) or 19 years (Hadcrut3). Whatever the measurement method, carbon dioxide emissions have been increasing and temperature has not. The greens' ideological model has failed the most elementary validation test.

Warmist and green media sycophants are pretty keen to show that the planet is warming due to human activities and pick narrow and recent time intervals to spruik their disaster predictions. I am reminded of the poem *Said Hanrahan* by John O'Brien in which the eponymous hero responds to all weather events with the cry: "We'll all be rooned."

In the last 10,500 years of the current interglacial, 9,099 were warmer than now. Some 6,000 to 4,500 years ago in the Holocene Maximum, it was warmer than at present and sea level was about 1.2 metres higher than at present. This was the peak of the interglacial that we now enjoy. It was only 8,000 years ago that there was no summer ice in the Arctic. Many solar scientists are now predicting that we are on the downhill run towards the next cold cycle, glaciation or ice age. I hope not. Some of us like it hot.

From the scientific perspective, if there is a claim that a certain year was the hottest year, then questions must be asked. What is the order of accuracy? Normally the order of accuracy is greater than the suggested temperature rise hence the claim is scientifically invalid. The suggested annual temperature rise for Australia of 0.17°C is less than the order of accuracy of many measuring stations used to deduce this trend because older measurements are ± 0.5°C. This is meant to be an average temperature rise but how does one construct an average with few measuring stations and even fewer in remote areas? This is mathematical nonsense.

How were measurements made?

Measurements have been made by many methods and many primary measurements are "corrected". Why are they corrected? Why do corrections show temperature rises yet primary data does not? How were the old measurements and new measurements integrated? Sometimes they are, sometimes they are not and sometimes they are "corrected". This does not give confidence. Was the distribution of measurements uneven? Most measurements are in cities, urban areas and airports and undergo a "correction" but there is no necessity to "correct" measurements in rural areas.

These speculations about the hottest year don't tell us about the natural thermostats such as precipitation and evaporation and ocean heat. They don't tell us about the past when there was far more carbon dioxide in the atmosphere than now yet there was no global warming as a result. They don't tell us about the changes in the main greenhouse gas in the atmosphere (water vapour). They don't tell us that we had a solar maximum coincidental with warming. They don't tell us that in 2013 there were fewer typhoons, more Arctic ice, more Antarctic ice, a decrease in sea surface temperature and that sea surface temperature drives atmospheric temperature, not the inverse.

While Australia is huffing and puffing about the warmest year, warmest summer, the angry summer or warmest month, the US National Climate Data Center in 2013 had more record low temperatures than measured before. Bitterly cold winters have occurred in the Northern Hemisphere over the last few years. Are these a result of human emissions of carbon dioxide that seem to produce both hot summers in Australia and perishingly cold winters in the Northern Hemisphere? Predicted winter droughts in the Northern Hemisphere were heralded by record rains and flooding.

Surely the regular droughts in Australia are due to increases in temperature? No. They are due to decreases in local rainfall. Australian rainfall is highly variable but has not decreased over the last 100 years, the longest measured drought in Australia lasted 69 years and, because there is no water to evaporate (and operate as an air cooler), air temperatures rise. Warmer temperatures do not create a drought but droughts create warmer temperatures. The Murray-Darling catchment in South-East Australia now contains three times as much water as was held naturally because of water management by dams and irrigation. This has drought-proofed the food bowl of Australia. Recent changes in management of the Murray-Darling catchment resulting from green pressure have destroyed much agriculture and sent many farmers broke. Well done greens. You've done your bit for the world.

Hiding embarrassing data

The UK Met. Office thought Christmas Eve 2011 would be a good day for burying not so good news. It was the graph the Met. Office didn't want you to see, in an episode which, according to one newspaper, represented "a crime against science and the public".

At that time bush fires set off by Australia's "hottest summer ever" were blamed on runaway global warming. A very large number of catastrophic bush fires are actually deliberately lit. Is this because global warming has made the brains of pyromaniacs soggy? If global warming is to be blamed for sporadic catastrophic bushfire events, why have *Eucalyptus* species evolved over millions of years to store their seeds in protected capsules only to release the seeds after a devastating fire? During the bushfire season in Australia, rather less attention was given to the heavy snow in Jerusalem (heaviest for 20 years) or the abnormal cold bringing death and destruction to China (worst for 30 years), and perishing cold weather in northern India (coldest for 77 years) and Alaska (with average temperatures down in the past decade by more than a degree).

But another story, which did attract coverage across the world, was the latest in a seemingly endless series of embarrassments for the UK Met. Office. The UK Met. Office sneaked on to its website a revised version of the graph it had posted a year earlier showing its prediction of global temperatures for the next five years. It was not until 5 January 2012 that sharp-eyed climate bloggers noticed how different this was from the one it replaced. When the two graphs were posted together on Tallbloke's Talkshop, this was soon picked up by the Global Warming Policy Foundation which whizzed it around the media.

The Met. Office's allies, such as the BBC, were soon trying to downplay the story, claiming that the forecast had only been revised by "a fifth", and that even if the temperature rise had temporarily stalled, due to "natural factors", the underlying warming trend would soon reappear. These rationalisations were made in an evidence-free zone. They were only able to get away with this by omitting to show the contrast between the two graphs. In 2011, the Met. Office's computer model prediction had shown temperatures over the next five years soaring to a level 0.8°C higher than their average between 1971 and 2000, and far higher than the previous record year, 1998. The new graph shows the lack of any significant warming for the past 15 years and this seems likely to continue. The standards of probity in the climate alarmist world are not those of the real world. In the corporate world, this would be called fraud.

The UK Met. Office has a central role in promoting the worldwide scare over global warming. The predictions of its computer models, through its alliance with the Climatic Research Unit at the University of East Anglia (centre of the Climategate emails fraud), have been accorded unique prestige by the UN's IPCC, ever since the global warming-obsessed John Houghton, then head of the Met. Office, played a key part in setting up the IPCC in 1988.

Back in 2007, the UK Labour Government was preparing the Climate Change Act to prevent human-induced global warming and the Met. Office made blatant attempts to influence political debate. The Climate Change Act was passed in the House of Commons 463 votes to 3 as snow was falling outside. Winters since then have been particularly bitter, perhaps a taste of what is to come. The Met. Office and their media champions, the BBC, have shown that they are not scientifically independent. Why should the UK taxpayer pay for scaremongering ideology at the expense of data? No heads rolled.

A major reason why the Met. Office's forecasts have come such croppers in recent years is that its computer models since 1990 have assumed that by far the most important influence on global temperature is the rise in atmospheric carbon dioxide. These models have ignored other factors and have not looked at the past. However, average temperature fell by 0.7°C in 2008 which is the same as the total temperature rise in the 20th century. Clearly something was wrong with the basics of the Met. Office models. It was probably because they did not take into account all the natural factors driving the climate, such as solar radiation, shifts in the major ocean currents and random El Niño events.

The Met. Office, by trying to hide data that was contrary to their ideology, showed that it is the major party in betraying the UK taxpayers and in creating the most expensive scare story the world has ever known. The Met. Office's data now shows that there has been more than 17 years without warming despite increased emissions of carbon dioxide. They have now made revised predictions suggesting that the "pause in warming" will continue until 2017. Don't hold your breath. Most of their previous long-term predictions have been hopelessly wrong. A recent editorial in *Nature Geoscience* tried to explain this pause in warming but only succeeded in showing that the science of climate change is not settled, that many known natural variables have never been used in computer models and that there are many aspects of natural science about which we know very little.

Antarctic and Arctic adventures

An attempt at climate alarmist niche tourism to Antarctica in January 2014 ended in farce. But it could have been a tragedy with multiple fatalities. It was promoted as an "expedition to answer questions about how climate change in the frozen continent might already be shifting weather patterns in Australia" by retracing the steps of Sir Douglas Mawson 100 years earlier. The tourists on this largely taxpayer-funded jaunt that cost $1.5 million found no flowers growing in meadows around Mawson's Hut in Antarctica and were not able to return as conquering heroes with the proof of human-induced global warming.

Chris Turney, plus wife and children, mustered paying tourists and a sympathetic free-loading media onto the *Akademik Shokalskiy* to watch with bated breath the heroic planet-saving scientists battling against the elements to measure the thinning ice to obtain their pre-ordained conclusion: Antarctic ice is melting. And it is all due to global warming and we sinners are the cause. Never mind the huge amount of liquid hydrocarbons burned to make these measurements and to get to Antarctica. The activist "scientists" ignored measurements made far more easily from satellites and history that show that the Antarctic ice sheet is currently expanding. This was monocular activism at its worst and was clearly not science.

However, nature has a sense of humour. The Russian gin palace was trapped in ice, a Chinese ice breaker sent to rescue these heroic adventurers also became trapped in ice, the real Antarctic research from bases was interrupted as an Australian ice breaker supply ship was diverted and the climate tourists were flown by a fossil fuel-driven helicopter to the warmth of a ship well away from the ice. The Americans, Australian and Chinese all ran up huge costs to rescue the passengers from the *Ship of Fools*. All sorts of excuses were invented to show that the climate science activists on the ship were not ill-prepared, incompetent, ignorant or aware of past ground and satellite measurements. When questioned about the failure, they resorted to obfuscation and dissimulation.

The climate activist community was silent, the normal suspects in the media became very creative with excuses (especially those on board) and the journal *Nature* showed that it was a magazine of political activism rather than one scientific independence. The expedition was to show that this area was warmer than when Sir Douglas Mawson was in the exact same place 100 years ago. The farce showed the exact opposite as, ironically, Mawson was able to penetrate much closer to land because of the lack of ice.

There was no chance of frostbite, eating huskies or death of companions, as Mawson experienced. It was all a bit of a giggle with a games program organised on the ice because passengers were getting bored. The rescue was conducted by vessels using fossil fuels, not wind or solar power. Dozens of tourist vessels visit the Antarctic without becoming trapped in ice. It appears that the only tourist ship ever to be trapped in ice in summer was one with climate "scientists" trying to show that the ice was disappearing.

If Chris Turney did not live off research grants and was not employed by a university, he would have been sacked for gross incompetence, breaches of safety protocols and misleading and deceptive conduct.

He was not and the taxpayer still keeps paying him. The public was not fooled, but activist climate "scientists" showed their true colours and even Turley's university defended him when they should have hidden him in a burrow.

As if this was not exciting enough, on the other side of the world the organisation that hangs banners from everything they can climb was active. Not only was the Antarctic freezing, the Northern Hemisphere was also.

Greenpeace activists were arrested by Russians for trying to stop Russians drilling for oil in Russian waters. What did Greenpeace expect? To try to climb onto a drilling rig in Arctic waters is against all safety protocols, endangered the lives of others and was a breach of Russian sovereignty. Does Greenpeace really think it is so important that it is above the law? Apparently so. There were government travel warnings that the activists ignored and, after arrest, those imprisoned called upon their governments to help them. Again, this cost the taxpayer a huge amount.

These two farces were wonderfully entertaining, helped the punter conclude that they will not fund the stupidity of others and showed the punter the shallowness of climate change activists' claims.

To paraphrase the ancient Greek writer Euripides: "Those whom the gods wish to destroy, they first make mad." Climate activist Chris Turney chose this path and those in his inner circle only provided him with encouragement.

Wild weather

Every day somewhere on Earth has wild weather. Years ago we only read about this later in newspapers. Now with modern communications systems and a highly competitive 24/7 news cycle, we learn of such wild weather as it is happening. In the past, cyclones, tornadoes, bushfires, rainstorms, snow storms, ice storms, floods and droughts were due to the weather. As communication systems improved and media and green groups needed to feed off disasters, this wild weather (now called extreme weather) was then deemed to be due to man-made global warming driven by man's sinful emissions of carbon dioxide.

Screaming babies, crying mothers, grieving fathers, concerned politicians and alarmists apportioning blame to others make for good television. Never let the truth get in the way of a good story. The truth just does not sell. There is a large body of evidence to show that we live in rather boring times, that the frequency of many of these events has actually decreased and that many past wild weather events were more severe than anything we view today in the lounge room on our televisions. With increased wealth and population, more people are building at waterside locations and more expensive residences elsewhere and hence the cost of property damage becomes higher.

From 1910-1940, many parts of the Earth were warmer than now. Temperature has been increasing for the last 300 years hence it is no surprise that the highest temperatures are towards the end of a warming period before we enter another solar-induced minimum. To suggest that the future will be hotter, drier and more vulnerable to floods ignores the past empirical evidence. The numerous recent predictions have all been wrong and there is no reason to believe that a repeated prediction will be correct. We ignore the past at our peril.

Ice on Earth is rare. For more than 80% of time, planet Earth has been warmer and wetter than at present and, since about 2,400 million years ago, the atmospheric carbon dioxide content has decreased from around 30% to the present 0.04%. The decrease in carbon dioxide results from the long-term biota-assisted sequestration into limey and carbon-rich rocks. There is a small amount of sequestration into solid rock when fluids and rock have chemically reacted.

Ocean acidification

In former times of high atmospheric carbon dioxide, oceans were not acid, there was no runaway greenhouse and the rate of change of temperature, sea level and ice waxing and waning was no different from the present. The alkalinity (measured as pH units on a logarithmic scale) of ocean water is slightly variable. A very slight change to ocean pH would involve a chemical reaction utilising monstrous volumes of acid. Seawater is not becoming acidified, it changes slightly in alkalinity. The lowest alkalinity (pH 7.3) is very close to acid hot springs. Any green, climate activist or journalist who refers to ocean acidity demonstrates a lack of knowledge of basic chemistry. Or maybe they are just deliberately misleading.

The oceans have been alkaline throughout the history of time because water chemistry, ocean floor sediments and new volcanic rocks on the sea floor buffer seawater to stop it becoming acid, even during times of carbon dioxide concentrations that were thousands of times the present value.

When we run out of rocks on the sea floor, the oceans will become acid. Don't wait up.

Sea levels

Coastal planning based on "global sea level rise" is asinine because other variables such as local compaction, sedimentation, uplift, erosion and subsidence are ignored. Not only does sea level change, the land level does also. And quickly. Computer-modelled sea level projections by the IPCC and governments have already been shown to be hopelessly wrong. For example, areas covered with ice sheets during the last glaciation (116,000-10,500 years ago) sank under the weight of ice. With the collapse of the ice sheets in the current interglacial, some lands are rising at present (e.g. Scandinavia, Scotland) and others are currently sinking (SE England, The Netherlands).

History shows us that some port cities (e.g. Efeses, Turkey) are now inland whereas other cities (e.g. Lydia, Turkey) are submerged. In both the Maldives and eastern Australia, relative sea level has fallen. The Maldives is 70 centimetres higher now than in the 1970s and eastern Australia is two metres higher than 4,000 years ago. Without a detailed knowledge of local land rises and falls, subsidence, erosion and sedimentation, global sea level predictions for coastal planning are only unfounded speculation.

Sea level changes are natural. Since the zenith of the last glaciation 20,000 years ago, sea level has risen 130 metres. What is expected after a glaciation? A sea level fall or a sea level rise? What is important is that the post-glacial rate of sea level rise is declining, exactly what would be expected at the end of an interglacial period. *Nature Geoscience* recently reported that since 2002, the rate of sea level rise has declined by 31%.

Charles Darwin showed in 1842 that as sea level rises, coral atolls grow and keep up with the sea level rise. His suggestion was that coral atolls growing on top of a volcano keep growing at a very rapid rate as the volcano subsides. The sinking of an island has the same effect as a sea level rise. It is a relative sea level rise. Darwin's theory was validated after drilling of coral atolls in the South Pacific Ocean in the late 19th and mid 20th centuries. His theory was again validated by drilling coral atoll in the Bahamas.

Elsewhere in the Pacific (e.g. Vanuatu), a local land level rise has elevated coral reefs and dead modern coral reefs occur well above sea level. If Pacific island nation states enjoyed a sea level rise, their land area would increase. This was suggested by Darwin and has been confirmed by recent satellite measurements. Compaction, use of coral for cement manufacture, roads and construction and extraction of ground water from unconsolidated coralline sand all lead to a relative sea level rise in the Pacific islands, as does polar ice cap melting.

Computer climate models

Computer climate models throw no new light on climate processes and the science underpinning the hypothesis that humans drive global warming is not in accord with the past. Climate models tell more about the modellers' ideology than the climate as they produce forward-modelled, pre-ordained conclusions but cannot be run backwards to show what has already been observed. Climate, atmospheric temperature and ocean temperature models have all been checked with empirical measurements and all models have been shown to be incorrect. Some 20 years ago, models showed us that atmospheric temperature would rise dramatically. Measurements over the last 20 years show that this has not been the case and, in essence, there has been no temperature rise. There is no fatal flaw with the measurements and the models therefore should be abandoned.

The media, climate "scientists" and computer modellers tell us that the science is settled. Where are all the scientific papers demonstrating model validation through ongoing reconciliations resulting in model refinement? There are none. If the science is settled then climate "scientists" have done their job and there is nothing more to do. Thank you very much. Now here's your retrenchment notice.

If one moves from Helsinki to Singapore, there is an average temperature increase of greater than 20°C yet this does not appear fatal. Humans have adapted to live on ice sheets, mountains, deserts, tropics and at sea level and can adapt to future changes. We live in the temperature range of -40°C to +50°C. Greens are telling us that a 2°C temperature change will be catastrophic. Pull the other one. We have such temperature changes hundreds of times each day (e.g. walking outside, changing rooms, diurnal temperatures, etc). Because of international travel, twice in 2014 I experienced temperature rises of more than 50°C. The jet lag and acclimatisation process was a temporary inconvenience but I think I am still alive.

Life on Earth has survived changes from glacial to interglacial conditions when temperature changes were far greater than 2°C. Life adapted. What the greens don't want to understand is that the planet is dynamic, that climate changes measured today are much smaller than past climate changes and that life just gets on with the business of being life and reproducing. Just because the planet changes in your lifetime does not necessarily mean that this is due to humans (unless of course hubris, narcissism, egocentricity and ignorance are seminal to your life). It is only young Earth creationists and greens who view planet Earth as static and the comparisons just don't end there.

History shows that during interglacials, humans created wealth and population grew whereas glacials are typified by famine, starvation, disease and depopulation. What would you prefer? We seem to have forgotten history. The peace and serenity we enjoy today were not the normal state of affairs.

The primacy of geology

The story of planet Earth is a marvellous chronicle written in stone. Past climate changes have been very complicated in a chaotic non-linear system with sporadic randomness: these systems are poorly understood and it is only by looking at the past and integrating with what we know about the present that we can hope to understand major natural processes. That understanding is a long way away. For scientists to argue that traces of a trace gas emitted by humans into the atmosphere are the main driving force for climate changes on planet Earth is fraudulent. To argue that every change on a dynamic planet is due to human activity ignores the rich past that the chronicle of planet Earth gives us. To throw off the past, we are eventually left with nothing and this is what the greens want. No past climate changes. Only a static planet whereby changes are due to our sinful carbon dioxide emissions.

Although knowledge of the history of planet Earth will always be incomplete, we have enough empirical evidence from the present, history, archaeology, glaciology and geology to show that past climate changes have never been driven by trace additions of carbon dioxide into the atmosphere; hence there is no reason to conclude that present human emissions of carbon dioxide will be any different. Carbon dioxide is not a pollutant. It is plant food. Without carbon dioxide, there would be no life on Earth.

Carbon dioxide is a minor greenhouse gas that has a very slight effect on atmospheric temperature. The human contribution to global carbon dioxide concentrations are, in essence, immaterial. The main greenhouse gas that affects atmospheric temperature is water vapour and, as the ice sheets show, the thermal properties of water explain why there has been no runaway global warming in the past when atmospheric carbon dioxide was far higher than now. Today's processes are tomorrow's geology. Yesterday's climate changes are preserved in rock outcrops that we can observe today.

Over the last 200 years of scientific activity, geology has been the ultimate climate science. It still is. This is why it is ignored by greens because it gives a chronicle based on evidence that is not in accord with their ideology. The greens and climate activists might ignore the past but it doesn't make it go away.

One often reads that geologists have a very different view of human-induced climate change compared to the view of global warming green catastrophists who make a living out of frightening us witless. This is because geologists are the ultimate climate scientists, particularly those employed in industry. Geologists are polymaths who take a broad view of the planet over time and continually test their ideas with methods such as drilling.

I hope you now see why geologists have a different view of the planet's climate.

The corruption of the scientific method

This history of planet Earth has been ignored with the current popular catastrophist paradigm of human-induced climate change. If large bodies of evidence and history are ignored, then this provides a misleading and deceptive view of global climate. If scientists ignore integrated interdisciplinary empirical evidence, then they have politicised science to gain government favours and they are operating fraudulently.

In *Objective Knowledge: An Evolutionary Approach*, the philosopher on the scientific method Karl Popper wrote: "Whenever a theory appears to you as the only possible one, take this as a sign that you have neither understood the theory nor the problem which it is intended to solve." This is why I argue that climate "scientists" are not scientists because they work with only one theory, their work has pre-ordained conclusions and they consider no other theories. Climate "scientists" are certainly green activists but not scientists.

Using Popperian thinking, if just one *iota* of validated evidence is not in accord with the hypothesis that human emissions of carbon dioxide drive global warming, then the hypothesis must be rejected. Geology provides numerous examples that falsify the hypothesis that human emissions drive climate change. Although the human-induced global warming hypothesis has been falsified by science many times, it is still cherished by greens, scientific activists and politicians for political, not scientific reasons.

Whether one is an expert or not in some minor area of atmospheric science, the popular view that human emissions of carbon dioxide change climate is not in accord with the scientific method because huge bodies of knowledge are ignored (e.g. astrophysics, solar physics, geology) and the theory of human-induced global warming is not in accord with what has been validated from other areas of science (i.e. coherence criterion of science).

Climategate and corruption

Climategate was a revelation. The adjustment of primary data to yield the required data for continuing the climate scare campaign, the corruption of the peer review process, the exclusion of contrary views from eminent scientists by the media, the lack of caution and reserve in making public statements about new scientific findings, the exaggeration of new scientific findings, the corruption of the temperature record, the non-correlation of carbon dioxide with temperature, the conversion of science and free independent inquiry into political advocacy, the corruption of the historical carbon dioxide record, the dampening or omission of the validated record of the Roman and Medieval Warmings, the creation of the "hockey stick" *ex nihilo*, the demonising of dissent, the denial that planet Earth changes by forces far larger than anything humans can create, the failure of models and predictions, the massive vested interests promoting the certainty of a human-induced catastrophe, the use of Orwellian language and the use of fundamentalist religious thinking processes all show that the gains made in the Renaissance have been lost in two short decades. Such behaviour above is not that of scientists. These are the tactics of paid political thugs. If the behaviour of the "scientists" involved in Climategate had taken place in the corporate world, then more gaols would have to be built.

The public has been deceived for a long time about human-induced climate change. They have now changed their mind, thanks to the vulgar behaviour of the climate "scientists" and the media operating as political advocates. And once people realise that they have been deceived and accordingly have changed their minds about human-induced climate change, they will not change their minds again.

The scientific hypothesis of human-induced global warming and the resulting "science" had pre-ordained conclusions and scientific facts were deliberately narrowed to deal only with carbon dioxide. Unaccountable government agencies and political pressure groups were involved and took control, scientific and political structures were put in place to enhance the deception. Actions were taken to block challenges. People's fears about change and catastrophe were exploited. The public's lack of understanding about the scientific method was exploited.

People find it hard to believe that deception on such a scale was possible. Those of us who were seen as opponents were viciously attacked causing others to remain silent. The attacks were *ad hominem* and the scientific arguments were never addressed. They still have not been addressed.

However, the public is not stupid and the constant braying of failed catastrophist doomsday scenarios by climate activists has had the opposite effect. Furthermore, nature has not co-operated over the intervening period with the catastrophist predictions made a few decades ago. On the other hand, there are a large number of otherwise unemployable climate "scientists", bureaucrats, politicians and green activists who have created a soft career for themselves by trying to frighten us witless about their forthcoming climate catastrophe and they will not go away despite the will of the electorate and the economic costs of their policies.

Much of the political and media pressure comes from full-time climate advocates paid to misinform. All this has been very damaging to science. When we really need science, perhaps in the next inevitable pandemic, the methodology of science may no longer exist and we'll have to resort to chanting and slapping ourselves with birch branches to avoid infectious disease.

What if I'm wrong?

But what if I am wrong and a reduction of carbon dioxide emissions is necessary to "save the planet"? What an odd question to ask? Greens and climate activists never ask "What if I am wrong?" This is a question that genuine scientists using Popperian thinking ask themselves.

If Australia reduced its carbon dioxide emissions by 5% by 2020, unvalidated models by climate "scientists" predict that there would be a cooling of between 0.0007° and 0.00007°C. This temperature decrease cannot be measured and such a restriction of emissions is pointless in the light of the great increase in emissions by the developing world. Nevertheless, I am sure that greens would feel good about reducing global temperature by such a "large" amount. Not only would Australia become bankrupt and be unable to feed itself, such voluntary acts of international environmental kindness would have absolutely no effect on the global climate. Such a self destructive sacrifice by Australia would not be reciprocated by developing nations such as China and India.

And if I am hopelessly wrong, then it is far easier to adapt to a hypothetical future climate change than to throw public money around in futile attempts now to prevent a hypothetical climate change.

Science is not blessed with certainty and the excitement of science is its uncertainty. Science has and will continue to self-correct as it has done many times in the past. However, Western governments have uncritically and dogmatically embraced human-induced global warming as a pretext for increasing taxation, redistributing wealth, eroding freedoms and for maintaining power by doing deals with unelected groups allegedly concerned about the environment while constraining liberal thinking processes. These changes take only decades to enact and centuries to reverse following massive economic and human disruption.

We frail humans commonly yield to fads, fashions, frauds and fools because we have short memories and ignore the past. We ignore social, political, economic and geologic history at our peril. This is happening now. To look at a moment of Earth history and use it to make predictions about the future is the folly of the greens. It is like watching five seconds of the love scene in the one hour 42 minute-long film *Casablanca* and concluding that the whole film is a love story. It is not. By the same token, to look at the last few years, decades or millennia gives an incorrect picture of the constant climate changes on planet Earth over the last 4,500 million years.

Spoons in antiquity

This book is actually about your stainless steel teaspoon so we had better move along. It is not really known whether the knife or the spoon came first. The earliest knives were made of bone and stone whereas the earliest spoons were made of wood and hence rotted over time. Spoons have been used for tens of thousands of years. Maybe the first spoon used was a large curved leaf, shell or gourd emulating a cupped human hand to get a drink of water from a stream. Prehistoric people used shells and wood for spoons. Ivory, bone, horn, pottery and stone were also used for spoons. Much later, metals such as gold, silver, bronze (initially copper-arsenic and later a copper-tin alloy), brass (copper-zinc alloy) and pewter (lead-tin-antimony alloy) were used.

Spoons have been known from all the past great cultures. The Romans designed a spoon for eating soft foods and soups and a smaller spoon for eating eggs and shellfish. We now have teaspoons, table spoons, dessert spoons, soup spoons, various spoons for the preparation of foods and dreadful plastic spoons. A spoonful is used as a poorly defined unit of measurement in food preparation. If you happen to end up at the wrong place at the wrong time, you might even hear spoons being played as a musical instrument.

In the Middle Ages in Europe and the UK, dinner guests bought their own wood or horn spoons. They often had their own spoon with them all the time, worn on their belt like a knife. The personal spoon was an item of display and was not carried for hygiene reasons. The wealthy and royalty, of course, had gold and silver spoons. Spoons of gold and silver are commonly hand forged and carried all sorts of markings and escutcheons. Some 500 years ago, a silver spoon was a method of displaying wealth hence expressions such as "born with a silver spoon". Silver from spoons would have acted as an anti-bacterial agent hence being wealthy was good for health (as it is today).

By the 14th century, spoons were made of tin-plated iron, brass, nickel-silver alloys, pewter and ceramics. Because pewter was malleable and used commonly available metals, it was used by the general public for spoons. Spoons made of brass, nickel-silver alloys and pewter all add poisonous trace elements to the body and all sorts of bacteria can live in the nooks and crannies of porous wood, shell, horn and bone spoons. The safest material to use for cutlery is stainless steel which was only discovered in the late 19th century. Infections from cutlery, cuts and tooth decay killed many before the 20th century.

Modern stainless steel teaspoons are cut from sheets of stainless steel, the bowl is cross-rolled and the shaft is rolled to the required length. This spoon is used as a metal template to shape a die that is used to shape additional spoons from cut slices of stainless steel. Stainless steel is widely used in kitchens, food transport and in cutlery because it is resistant to stains, rust, accumulation of bacteria and corrosion.

Besides using stainless steel in everyday life for food handling, storing, cooking, serving and cutlery, stainless steel has numerous other uses. Drinks such as fruit juice, beer, milk, wine and soft drinks are processed in stainless steel equipment. Stainless steel is also used in commercial cookers, pasteurisers, transfer bins and wine fermenters. This is because while being easily cleaned, corrosion resistant, durable and cheap, stainless steel protects food flavours. Porous material such as wooden spoons suck up flavours, can have a city of bacteria living in the pores and are not very durable.

The European 19th century fad of collecting souvenir spoons has infected most Western cultures. Such spoons are to celebrate all sorts of anniversaries and to boast about the unspeakable places visited on holidays. There is now quite a collectors market in souvenir spoons.

Who really did invent stainless steel?

There is quite a battle for discovery rights. Was it the English, French, Germans or Americans? It seems that stainless steel was the end result of a long process of experimenting by many people with all sorts of alloys. This could only be done when steel could be manufactured to consistent quality commercially by the Bessemer Process.

The 19th century was a period of great experimentation and the blossoming of the arts, science and engineering. When large amounts of pig iron could be cheaply made, there was much experimentation because it was very brittle. In 1820, Michael Faraday found that nickel-bearing iron was far stronger than pig or wrought iron. One of the experiments by Pierre Berthier in 1821 noted that iron-chromium alloys were resistant to attack by acids and Berthier suggested that such alloys could be used in cutlery. However, because these iron-chromium alloys had a high carbon content, they were too brittle to be useful. Incidentally, an iron antimony sulphide mineral called berthierite was discovered in 1827 and named after him. All the best people have a mineral named after them.

The thermite process has been used for a long time for welding, especially in areas with little infrastructure. By lighting very finely powdered aluminium metal and iron oxide, a huge amount of heat is created, the aluminium strips the oxygen from the iron oxide and the residue is molten iron and aluminium oxide. Welding of railway lines in remote areas was commonly by the thermite method (e.g. the Nullarbor transcontinental east-west railway in Australia). The same method was used in the late 19th century to produce carbon-free chromium using finely powdered aluminium and chromium oxide. Any traces of carbon are quickly burned. If chromite is used, ferrochrome can be produced.

The French scientist Leon Gillet documented the composition and properties of an alloy similar to stainless steel. He never realised the corrosion resistance of his alloy. Although the English metallurgist Harry Brearley sometimes gets the credit for the discovery of stainless steel, Leon Gillet is also a candidate as are a few others such as Giesen, Portevin, Monnartz, Borchers and Mauermann.

In the early 20th century, metallurgists were able to create stainless steel and the relationship between the chromium content of steel and corrosion resistance was determined. In 1908, Krupps built a chromium-nickel hull for the 370 tonne yacht *Germania*. Two Krupp engineers patented very low carbon stainless steel in 1912. High-carbon high chromium stainless steels were being made on an industrial scale in the USA and in 1912 a patent was applied for a high chromium 1% carbon steel alloy by Elwood Haynes.

Concurrently, a high chromium 1% carbon steel was invented in the UK in 1912 by Harry Brearley at the steel city of Sheffield. He was looking for corrosion resistant alloys for gun barrels. Brearley conducted tests to create new alloys. He noticed that the discarded sample from an earlier test had not rusted yet other samples had. On applying for a US patent, Brearley found that he had been pipped at the post by Haynes, so they pooled their resources and formed the American Stainless Steel Corporation in Pittsburgh. Haynes' patent was granted in 1919.

How is stainless steel made?

To make stainless steel now, the raw materials are melted together for 8 to 12 hours in an electric arc furnace. A mixture of low-carbon steel, ferrochrome, ferronickel, nickel and maybe molybdenum and tin are melted together. When the melt is homogeneous the liquid alloy is cast into blooms, billets, rods, tube, rounds and slabs.

Blooms and billets are formed into bar and wire, slabs are rolled into plate, strip and sheet. Bars come in all grades of stainless steel as rounds, squares, octagons and hexagons. Stainless steel then goes through a process of heating and cooling under controlled conditions that reduces internal stresses, and softens and strengthens the alloy. This annealing process may involve up to two hours of quenching hot stainless steel in an ice-water bath or quenching with an air blast. If stainless steel is cooled slowly, undesirable materials (carbides) form in the alloy. Hot rolling, annealing and descaling give a dull finish whereas cold rolling produces a better finish. Annealing causes scale to form on the steel surface. It is removed by pickling in a nitric acid-hydrofluoric acid bath or electro cleaning in a phosphoric acid solution where stainless steel is the cathode.

Stainless steel is now ready for producing the required shapes of the blanks. Cutting is by mechanical methods such as shearing with guillotine knives, sawing, punching in dies or drilling overlapping holes. More expensive cutting is by flame cutting using an oxygen-propane mixture with iron powder or by plasma jet cutting. The stainless steel is then polished to give a finish and to allow a new oxide layer to form. A mirror finish is produced by grinding with progressively finer abrasive followed by extensive buffing. The abrasives used are minerals such as diamond, garnet (iron aluminium silicate), corundum (aluminium oxide) or emery (magnetite-corundum mix) or abrasives manufactured from other minerals such as cerium, lanthanum and tin oxides. The polished stainless steel is now packed and transported for fabrication. The steel is rolled, pressed, forged, drawn and extruded and may have to undergo further heat treating, machining and cleaning. Larger structures are welded by fusion using an electric arc between an electrode and the stainless steel.

Stainless steel teaspoons can be purchased from the fabricators in units of 1,000 to 20,000 at a cost of $0.01 to $2.00 each, depending upon quality, design and robustness. The alloy composition is commonly stamped on the back of a stainless steel teaspoon. Better quality stainless steel cutlery is 18:10 (18% chromium, 10% nickel) whereas most cutlery is 18:8. No matter where you live, the retail price will be considerably higher. The amount of effort and energy to produce such a complex useful alloy just does not seem commensurate with the price. But where once you needed to be royal or rich to be "born with a silver spoon", now we can all be born with a stainless steel spoon and be the better for it.

Stainless steel

Stainless steel is principally an alloy of iron, chromium and nickel, the chromium being the key. At times, molybdenum, manganese, tin or other metals are added. Chromium in nature very quickly bonds firmly with oxygen and this high affinity of chromium for oxygen allows the stainless steel alloy to form a stable extremely thin chromium oxide film at the surface that stops corrosion. The chromium oxide film is impervious to water and air hence protects everything covered by it. The film is so thin that it does not affect the lustre of stainless steel. This film is called the passive oxide layer (passivation layer) and it forms instantaneously when the stainless steel is exposed to the atmosphere. All metals, except gold, platinum and palladium, corrode spontaneously when in contact with air. Stainless steel is also self-healing and rebuilds when the oxide layer has been removed. The nickel preserves the internal structure of steel. A cheaper metal manganese can also do the same job but not as well.

Every day we use the corrosion resistance of stainless steel for food preparation and eating. Many buildings, fences and memorials are clad by or composed of stainless steel because it is very resistant to the elements. Stainless steel is strong, ductile, long-lasting, inexpensive (compared to other metals) and can easily be re-melted and recycled. About 60% of the stainless steel used is recycled. However, most recycled stainless steel is used for industrial applications with 60% left over from manufacturing and 40% from end of working life industrial use. Your humble stainless steel teaspoon at the end of its life is generally not recycled.

Because there are many uses for stainless steel, there are about 150 different grades of the alloy, of which 15 are most common. The different grades reflect the different chemistry of the alloys. Stainless steel alloys are milled into coils, sheets, plates, bars, wires and tubing for manufacture into special appliances and construction materials on large buildings. The average person in a Western country uses 80 to 180 kilograms of stainless steel in their life whereas in the developing world it is about 15 kilograms.

Austenitic stainless steel (also known as the 300 series) is the most common and most widely used stainless steel. It can be cold worked and welded and becomes slightly magnetic after cold working. When heated and cooled many times, it loses its magnetism. Various specific varieties of 300 series stainless steels have added sulphur, manganese, lead, titanium, selenium and molybdenum. T304 is 18:8 stainless steel

(18% chromium, 8% nickel, 0.15% carbon) and is the most common alloy used for cutlery. It is ductile and highly resistant to corrosion. Many spoon benders (such as the author, Uri Geller *et al.*) are well aware of both the ductile and brittle behaviour of stainless steel and use these properties to demonstrate the spoon bender's supernatural powers. Or maybe we demonstrate that audiences are gullible and easily conned? If people can so easily be conned by spoon bending then they also can be conned with human-induced global warming.

Type 316 stainless steel is used to cover large buildings (e.g. Petronas Twin Towers, Kuala Lumpur), 302 stainless steel covers the Chrysler Building (New York) and Australia's Federal parliament has a 220 tonne flagpole made of 304 stainless steel. Some 300 series steels are used in medical implants because of their biocompatibility (e.g. 316LMV).

A stainless steel with a higher carbon content (1%), 18% chromium, nickel and molybdenum is martensitic or 400 series stainless steel. It can be heat hardened, can be magnetic and is difficult to weld because of the high carbon content. It is less durable than 18:8 stainless steel but more corrosion resistant so it is used in surgical equipment, better quality cutlery and for making moulds. The 409 series stainless steel is used for car exhaust systems because it is both corrosion and heat resistant. It also looks good to the average rev head, such as the author.

Another 400 series stainless steel is a ferritic alloy containing high chromium. The 440C alloy contains 16 to 18% chromium and 0.95 to 1.2% carbon whereas the 446 alloy contains 23 to 27% chromium and a maximum of 0.35% carbon. Although less ductile than other varieties and cannot be hardened by heating, the 400 series is less durable and less tough than 300 series stainless steel. However, it is extremely highly resistant to corrosion and is used in boats, boilers and washing machines. Molybdenum is added to stainless steel for use in extremely corrosive environments where mechanical strength is required (e.g. waterside construction).

Other stainless steel with even more chromium is used in very corrosive high stress environments such as high temperature-high pressure autoclaves, high chloride environments (e.g. desalination plants), heat exchangers, chemical tankers, chemical reactor vessels, flue gas filters, corrosive liquid distillation (e.g. acetic acid), petrochemical plants, offshore rigs and various oil and gas equipment. Other stainless steels harden at a relatively low temperature, they have small quantities of molybdenum, tin, copper, niobium and aluminium, and are used for pumps, engine shafts and aeroplane components.

The human chain

Not a single person on Earth could make a stainless steel teaspoon all by him or herself. No one person can mine the commodities required, smelt the metals, transport the metals and alloys and fabricate a stainless steel teaspoon. We are dependent upon numerous heavy industries to give us the simple stainless steel teaspoon. No country contains the spectrum of natural resources necessary to make a stainless steel teaspoon hence it can only be made if there is international trade. The construction of a simple stainless steel teaspoon derives from the efforts of thousands of people. These people must provide the food, housing, electricity, water, clothing and transport for mine, smelter and factory workers. This is an extraordinary logistical achievement.

Little does an African corn grower providing food to chrome miners know that his farming is part of a process that leads to the making of a stainless steel teaspoon to feed people on the other side of the world. A stainless steel teaspoon that he also would use after a hard day of toiling in the fields. And then there are those that are part of the process to make clothes for the miner. And on it goes. These separate unconnected people have never met yet, through a price system, they come together to make something very complex. A stainless steel teaspoon.

As soon as there is a better way of mining chrome, coal, iron, nickel, molybdenum and tin needed for a stainless steel teaspoon, it will be done to cut costs. As soon as there are more efficient ways of transport or better manufacturing procedures and technology, they will be used somewhere in the world. This is how entrepreneurs break business models, monopolies and old technology.

The world is highly competitive, not conspiratorial. Conspiracy theorists are paranoid. Someone would break ranks very quickly if there were a conspiracy to suppress an invention that was cheaper or more environmentally friendly. This would be done for financial gain, not for the honour of maintaining a conspiracy. There is an army waiting outside the gates of Spoon City trying to make their part of the process easier, cheaper and more profitable.

Our planetary fingerprint

Using 2011 US figures below, a considerable tonnage of minerals must be provided for every person in the United States to make items used every day. Australian, UK and EU figures would be roughly comparable. Just to make roads, buildings and bridges, for landscaping and for numerous chemical and construction applications, each American uses 8,509 lbs (3,799 kg) of stone. This needs to be drilled, blasted, loaded, transported, crushed, sized, lifted, loaded and transported. This is the mining most do not see and these processes use energy and the latest technology.

On average, each American is currently using annually 6,792 lbs (3,032 kg) of coal. Coal is the backbone of electricity generation and metals manufacture. Another form of energy is gas (80.905 cu ft per person *per annum*) and it is used in the chemicals industry, smelting, electricity generation, transportation and domestic uses. In addition to the coal and gas, 951 gallons (3,804 litres) of crude oil is used for transport and creating chemicals. Only 0.25 lb (0.12 kg) of uranium is used by every American each year for electricity generation by nuclear reactors.

Some 5,599 lbs (2,500 kg) of sand and gravel per person are used to make concrete, asphalt for roads, building blocks and bricks. These are commodities we could not do without. We cannot survive in the modern world without cement. Every person in the US uses 496 lbs (221 kg) of cement each year. To make this cement, limestone needs to be burned to lime and mixed with burned shale. This process releases huge amounts of carbon dioxide because limestone contains 44% carbon dioxide. Each person who uses anything made of concrete or cement, is responsible for putting great quantities of carbon dioxide into the atmosphere. The very action of greens walking on a concrete footpath or cycling down a road means that they are participants in emitting carbon dioxide into the atmosphere.

In the USA, salt is used for highway de-icing with smaller amounts used for chemicals, food, glass and agriculture and each US citizen uses 421 lb (188 kg) of salt each year.

Iron ore is used to make steel that is used for buildings, motor vehicles, railways and construction. To convert 357 lbs (162 kg) *per annum* of iron ore into steel requires a huge amount of energy and the release of carbon dioxide into the atmosphere.

Many other non-metallic minerals are mined for numerous uses such as boron for glass, gypsum for building board and agriculture, talc for cosmetics and calcite for paper. Other uses are for glass, chemicals, soaps, computers, medicines and cell phones. A total of 332 lbs (148 kg) of other non-metallic minerals is used each year by every person in the USA.

We need mining for dining and we could not feed ourselves without phosphate and potash. Phosphate rock is reacted with acid to form superphosphate fertiliser. This acid is normally a by-product from smelters that emit carbon dioxide into the atmosphere. Superphosphate slowly releases phosphate into soils and phosphorus is an essential nutrient for plant growth. Animal feed supplements also use phosphate. Some 217 lbs of phosphate rock (97 kg) is used each year by each American.

Clays are not really very sexy. However, each year each American uses 164 lbs (73 kg) of clays in floor and wall tiles, crockery, kitty litter, bricks, cement, paints, pencils and paper. Many junk foods contain clays. Clay poultices have been used for thousands of years in medicine and some clays (e.g. bentonite) are wonderful for cleaning up pollutants.

Aluminium metal is the final product from the mining of bauxite and on average 65 lbs (29 kg) is used each year by each American for housing, drink containers, cars, foil and aeroplanes. Although it does not sound very exciting, soda ash is made from salt. Some 36 lbs (16 kg) is used by each American each year to make glass, detergents, medicines and food additives and is used in photography and water treatment.

We each use 24 lbs (11 kg) each year of other metals for alloys, electronics and chemicals. For example, tungsten is used for hardening steel, creating tungsten carbide for drilling and cutting bits and for light bulb filaments. The modern would not survive without tungsten. Copper (12 lbs or 5.4 kg *per annum* per person) is used in buildings, electrical and electronic parts and transport. Another metal used for transport is lead in vehicle batteries (87% use) with minor uses in electrical, communications and computer/television screens (11 lbs or 4.9 kg per person *per annum*). There is a move in the EU to use less lead in electrical equipment for fear of toxic accumulations and solder (lead-tin alloy) is being replaced by other alloys of tin.

The main use of zinc (6 lbs or 2.7 kg per person *per annum*) is as a sacrificial metal for the galvanising of steel to create rust resistance. Zinc is also used in alloys such as brass and in paint, rubber, skin creams, health care and nutrition. Zinc chromate paints protect steel from rusting. Vehicle tyres contain zinc oxide which prevents rubber decomposition by sunlight. Zinc oxide in sunscreens absorbs UV radiation. Each person uses 5 lbs (2.2 kg) of manganese each year for steels and batteries. Lithium, tin, bismuth, cobalt, rare earth elements and many other elements are mined for speciality uses and these uses are always changing in a highly creative and competitive world.

We use about 180 tonnes of water per person *per annum*, mainly for industrial processes. Much of the industrial water is recycled, most domestic water is not. In Holland, domestic water comes from the Rhine River and it has already been through the human body a number of times since precipitation as snow and rain in the Alps. Some of the materials we use can be recycled (e.g. iron, copper, lead, aluminium, glass) but most of the materials we use are not (e.g. sand, gravel, cement, salt, non-metallics, fertilisers, etc). Recycling is only possible for certain commodities in large population centres.

About 100 years ago, the average Westerner used only 27 elements in the periodic table of 92 natural elements. Now the average modern Westerner uses more than 60 elements. Furthermore, the total tonnage of commodities used has increased ten-fold over the last 100 years.

There is no shortage of commodities. Uses change with availability, demand, prices and new inventions and it is very hard for a corporation or country to have a monopoly of a commodity for a long time. A very large number of discovered marginal mineral deposits are in the starting blocks waiting for price, supply and demand and political changes and if a commodity shortage drives up prices, these deposits can quickly provide commodities to the markets.

Furthermore, the fear that we will run out of a commodity is illogical. The law of supply and demand ensures it is almost impossible to run out of anything. As the stock of any commodity is depleted, the price rises. If there are no new mineral discoveries and the commodity becomes extremely rare, the price will rise so high that the commodity will become unaffordable. To anyone. Investment will then be directed into developing technologies utilising substitutes. The fear is irrational.

We're all going to die

The carbon dioxide in the atmosphere ends up in the oceans as dissolved carbon dioxide, bicarbonate and carbonate or as vegetation, limey sediments and fossils fuels. We humans do not emit carbon, we emit carbon dioxide. They are two very different things and any talk of carbon emissions is a display of ignorance. This has been perfected by greens and politicians who were once lawyers or social workers.

Marine creatures use these carbon compounds to construct calcium carbonate shells and it is this material that is dissolved back into ocean water at depth. In shallow waters, the shells accumulate as shell banks, carbonate-bearing sediments or in coral reefs. Waters moving through porous and permeable sediments cement grains together with carbonate minerals thereby sequestering carbon dioxide for up to hundreds of millions of years. When slow bottom-hugging currents eventually bring deep cold water to the surface, carbon dioxide is released to the atmosphere.

This creates a difficult problem for climate catastrophists who claim that human emissions of carbon dioxide are going to make us fry and die. The annual emissions of carbon dioxide by humans is 3% of the total. The other 97% is from ocean degassing, animal exhalation, volcanoes and natural oxidation. It has yet to be shown by the greens that it is the human emissions of carbon dioxide that drive climate change whereas the far larger natural emissions do not.

Again I must stress that in the atmosphere, carbon dioxide is a trace gas (400 parts per million by volume) and human emissions are adding traces of a trace gas to it. We are told by greens that human emissions of carbon dioxide create global warming, that this leads to climate change, extreme weather and all sorts of catastrophic events, that there will eventually be a tipping point and that the climate change will be catastrophic. The argument then regresses to suggesting that human emissions of carbon dioxide should be reduced. The end result would put the world back into the 18th century with increased mortality, increased unemployment and increased misery.

Is it any wonder that many rational people regard the greens as anti-human? The end result of green ideology is an attack on the coal, gas, oil and electricity industries and the forcing of expensive inefficient subsidised wind, solar, biomass, wave and tidal power onto the taxpayer. Rather than supporting sustainable industries such as forestry, fishing and farming, these are also attacked by the greens. It appears that anyone who is not sucking off the public teat is a target for attack by the greens.

In effect, the greens use carbon dioxide as a proxy for attacking industry, progress and the modern world. It has nothing to do with climate change, the environment or science. I am quite happy with the world today, if the greens want to go back hundreds of years then they can lead by example and go back to a primitive life. But don't expect all of us to join you. And come the next inevitable famine, don't come knocking on my door.

The future is bright

The modern world is amazing compared with what mankind experienced in the past. And the future is looking even brighter. Life expectancy has risen by about 30% in my lifetime, child mortality has fallen by two-thirds and income per person in real terms has trebled. More people are leaving poverty than in any time in human history. The average IQ is increasing in most countries, participation in education is increasing, more people live in democracies than previously and people in poor countries are getting rich faster than those in wealthy countries.

The large gap between global rich and poor is closing quickly. The death rate from extreme weather such as floods and storms has fallen by 98% in the last century. In 1800, six hours work on the average wage produced enough money to buy a candle that would burn for an hour. Today only half a second's average wage work is enough to buy electricity to keep a light on for an hour.

The golden age will be the future and to think of a past mythical golden age is just misty-eyed dreaming. For thousands of years we have been plagued by the prophets of doom predicting the end of the world. If just one of these prophets were correct, we would not be here. Most of these predictions were false and the rest of them were grossly and irresponsibly exaggerated. The greens are the modern day prophets of doom.

However, many in the West are totally unaware of the great advances made and our bright future. We are now so wealthy that we can afford to switch off the lights for an hour and step back into the Dark Ages to raise awareness about mythical climate change. Earth Hour does not occur in Asia, Africa and most of South America. In the West, the Earth Hour celebration of darkness is one big self-indulgent party. Many in the Third World would pray to have electricity for an hour. North Korea is in a permanent Earth Hour. Earth Hour does nothing whatsoever about reducing carbon dioxide emissions. In fact it does the opposite. It is a misguided preference for feeling good over doing good. Coal-fired generating systems do not shut down during Earth Hour when most participants turn off the lights and burn a candle.

Candles are about 300 times less efficient that fluorescent lights and emit carbon dioxide and dangerous unburned carbon compounds. Although a light here or there around the world is extinguished for an hour, all other electrical equipment still works. If the pagan darkness worshippers were serious, all electrical systems would be shut down including TV, radio, heating, cooling, computers, kitchen equipment, transport, hospitals and so on.

The billion or so well-intentioned people around the planet who participate in Earth Hour totally ignore that more than 1.3 billion people around the globe continue to live without electricity. These poor folk don't even have the option of turning off the lights for an hour. An estimated 3.5 billion people in the Third World burn dung and twigs indoors to cook and keep warm. Burning releases noxious fumes that kill an estimated 3.5 million people a year, mainly women and children. A simple electric stove would solve this problem by using reticulated electricity generated from coal.

It is immoral for the greens to keep the death rate high in the Third

World rather than advocating for efficient low-cost coal-fired electricity.

Merchants of doom

In my lifetime I have been told that global population growth was unstoppable and that mass famine and social disruption would result. It didn't. Apparently we were going to suffer a cancer epidemic because of the increased use of chemicals that would reside in the environment and shorten our lives. This didn't happen. We spent decades waiting around to die from the effects of DDT that was used widely when I was a child. We didn't die. In those malaria stricken areas that were not sprayed with DDT, people died. They still do, especially children. We were going to glow in the night from radioactivity. We didn't. There was an imminent nuclear holocaust. It did not happen. The Sahara Desert was expanding kilometres a year, this was our fault and North Africa would be inundated by sand. It was not.

We were going to run out of oil, we were going to run out of minerals, we were going to run out of water and we were going to run out of food. These did not happen. Acid rain was going to convert industrial nations into deforested deserts, we were going to fry and die because of the loss of the ozone hole, sperm counts were decreasing and humans were on the path to extinction, all sorts of viruses and bacteria were going to wipe us out and every day of our lives we seemed to be at a turning point in history. None of this happened. The very same people who in the 1970s were telling us that we are going to freeze in a forthcoming human-induced ice age are now telling us that we will die and fry from human-induced global warming. We didn't freeze and die. We also don't forget the climate scare stories of the 1970s.

At the turn of the millennium, we were told that computer systems would shut down, utilities would collapse and planes would drop out of the sky. The change from the old millennium to the new millennium was one big yawn. Those who sold software and serviced computers did very well, thank you very much.

Now we are again told that we are going to fry and die, that it's all our fault and a sea level rise will inundate coastal areas. Again, all our fault. The IPCC issued tipping point warnings in 1982, another 10-year tipping point in 1989. In 1982, UN official Mostafa Tolba, executive director of the UN Environment Program declared that the "world faces an ecological disaster as final as nuclear war within a couple of decades unless governments now act".

There was something strange in the water all over the world in 2009. In January 2009, NASA's James Hansen, author of the fraudulent "hockey stick" declared that "President Obama has four years to save the Earth". This is no mean feat during the first term of a President. The four years since January 2009 has passed. The Earth has been saved from whatever was going to destroy it but not by President Obama, greens or crippling green legislation. In May 2009, Elizabeth May of the Canadian Greens Party declared that "we have hours" to prevent a climate disaster. Not to be outdone, the UK's Gordon Brown warned of a global warming catastrophe and that there are only "50 days to save the world". In July 2009, Prince Charles claimed a 96-month tipping point. In October 2009, the WWF stated that that there were only "five years to save the world". The world has been saved from whatever was going to wipe it out by ignoring the WWF.

The media has a field day with predictions and even join in the fun of being consistently wrong. In April 2014, the *Boston Globe* noted: "The world now has a rough deadline for action on climate change. Nations need to take aggressive action in the next 15 years to cut carbon emissions, in order to forestall the worst effects of global warming, says the IPCC." Another big yawn. We've heard it all before. I am happy to be on the side of history and state that this future global warming disaster scenario also will not happen. The voice of reason was not heard. In 2007, a prominent New Zealand scientist claimed regarding global warming: "It's all going to be a joke in 5 years". This has shown to be correct. Another gong to the little country that punches well above its weight.

Poverty, ill health, environmental degradation and illiteracy are all decreasing. The world is a better place than before. This is shown in the global GDP when at the beginning of the 20th century the cost of poor health was 32% of global GDP whereas it is now 11%. This is despite population increase. Even a slight addition of carbon dioxide to the atmosphere has greened the planet resulting in more food production and a decrease in land clearing. Yet the greens try to tell us that carbon dioxide is a pollutant when it is in fact plant food. If the greens really wanted to green the planet, they would be encouraging industry and individuals to emit as much carbon dioxide as possible. The greens knowingly try to obfuscate by talking of carbon pollution. Carbon is a black solid and, by contrast, carbon dioxide is the colourless, odourless, tasteless gas of life. Without carbon dioxide, no plants. Without plants, no animals. Without carbon dioxide, no humans.

However, what about the other 97% of emissions? In green thinking, it appears that only human emissions of carbon dioxide drive climate change and that natural emissions do not. Furthermore, a very slight change in ocean degassing easily accounts for increases in atmospheric carbon dioxide. This would happen if the oceans slightly warmed. This occurs during warmer interglacials between two events of glaciation well after the interglacial warming starts or when there was a slight increase in sea floor global volcanicity a few thousand years ago. This is exactly where the planet is in its history in the current ice age. It has yet to be shown that the very slight increase in atmospheric carbon dioxide is unequivocally due to human activity. I have been waiting 30 years for this evidence. I am patient but after 30 years I think it can be safely concluded that the human-induced global warming scare campaign was a deliberately deceptive political campaign by greens, activist "scientists" and the lunar left. As a result, the penthouse proletariat profits and the punter pays.

As countries become wealthy, they can afford to undertake massive environmental programs. It is ironical that once countries become wealthy, common sense disappears and groups such as the greens rise and are actually listened to because some feel guilty that they live in a good geographic part of the world, have a stable democratic government and are relatively wealthy. The end result of green policies would be to create poor countries where there is just not enough money for environmental programs, let alone defence, education and health.

Now, back to our humble stainless steel teaspoon. There is no way we could make a stainless steel teaspoon without stable base load power. Well, what about "renewable" energy? Can't we just use "renewable" energy to keep the wheels of modern industry spinning?

2

ENERGY

How much energy do we need to make a stainless steel teaspoon and deliver it to your house? Is there a more efficient and lower cost way of making a stainless steel teaspoon? Is there a more environmentally friendly way of making a stainless steel teaspoon?

The whole process of exploration, mining, smelting, manufacture, distribution and retail sale of a stainless steel spoon is dependent upon energy. Large amounts of energy. This energy needs to be cheap, efficient, reliable and continuous. This is energy that is effectively embedded in the spoon.

To mine rocks we need energy. For our stainless steel teaspoon, we need to mine iron, chromium and nickel and possibly molybdenum, tin and other ores. If, for some reason, we decide to stop mining nickel or impose export restrictions in one country, then another country will mine the nickel required for the market. This has happened before and will happen again.

Mining is an international and highly competitive industry. The biggest cost in a mining operation is energy for ore haulage, crushing, grinding and selective separation and beneficiation and transport of the product. Some 2% of the world's energy is used for crushing and grinding rocks. Most of this energy comes from coal. No amount of wind, solar, tidal or biomass energy could provide the electricity and liquid fuels needed for a non-stop mining operation. If a mining operation is stopped until the wind blows or the sun shines for electricity, then it becomes unviable.

There are always metal losses when an operation shuts down and then starts up, such as for scheduled routine maintenance and unreliable electricity sources would only exacerbate an existing problem. Engineers have been working on these losses for centuries and I invite any green to use knowledge and creativity to reduce these losses that result in a saving of energy and more efficiency.

Once ores are mined, beneficiated or concentrated, the finished product needs to be transported to a smelter. Trucks, trains and ships are the normal transport methods. These use energy. Again, no ship at sea with thousands of tonnes of mineral cargo could use wind, solar or ocean power. There are now purpose-built ships that carry 400,000 tonnes of iron ore from Brazil to China. There is just not enough grunt (energy density) in the wind and Sun to move such a load. Of course, more energy is used for the loading and unloading of the product.

Concentrates need to be converted from oxides or sulphides to metals that are used to make the stainless steel alloy. This is a process that involves energy, oxidation, and reduction processes. A large amount of energy is needed for a smelter and especially a refinery. A smelter cannot be turned off and on at will otherwise the liquid rock in the smelter solidifies and it is a time-consuming and costly process to remove the solid material, repair the furnace lining and start again. Smelting is a continuous process with rare scheduled shutdowns for maintenance and replacement of the furnace lining.

The process for smelting nickel sulphide concentrates for your spoon is different from the process used to make iron. After sulphide mineral concentrate oxidation by roasting, oxides need to be chemically reduced. Iron ore needs to be reduced in order to remove oxygen from iron oxide to make iron metal. This process involves adding a reducing agent that generates heat such as coking coal, coke or gas. No amount of chemical wizardry can be used to force wind- or solar-generated electricity to perform the chemical process of reduction. A chemical reaction using coal, carbon as charcoal, carbon monoxide or hydrogen, is needed for chemical reduction. Unless, of course, the greens change the fundamentals of chemistry.

Once metals are produced by smelting, they are then fabricated into a stainless steel teaspoon. This uses energy. Large amounts of energy are needed continuously. No wind or solar energy can keep such a factory in constant production. No stainless steel teaspoon fabrication factory can have workers waiting around until the wind blows or the Sun shines. Once the spoon has been manufactured, then it needs to be transported and handled many times before it ends up in your kitchen. This international trade requires energy for transport.

Could we create the stainless steel teaspoon that you use to eat with from "renewable" energy? Let's look at it.

Energy density

Western greens have a penchant for self-destruction and advocating technological solutions that failed hundreds of years ago and not critically analysing their own solutions to perceived environmental problems. They amusingly call themselves "progressive", yet their track record shows they are highly retrogressive. The best example of this is green lip-flapping about energy.

To make 1,000 kilowatt hours of electricity requires one of the following:

- 0.15 grams of enriched uranium;
- 264.5 litres of oil;
- 379 kilograms of black coal;
- 1,000 one metre square solar panels operating for one day; or
- one 660-kilowatt wind turbine operating flat out for 1.5 hours.

It is clear that uranium, oil and coal contain the largest amount of energy per unit weight and that solar and wind energy are pretty unimpressive. If we are really interested in conserving resources, efficiency and saving the planet (from God knows what), then it is a no brainer. We should be generating electricity from uranium fission. For all of my life, nuclear fusion has only been 20 years away from utilisation. It still is.

Various solid and liquid fuels release variable amounts of energy (kJ, kilojoules) per unit weight (kg, kilograms) and are used for transport and electricity generation. These figures assume that all the heat used to convert the large volume of water in brown coal, peat and black coal to steam is recovered by steam condensation. However, much this heat is not recovered.

Examples are:

Gas (g)/liquid (l)/solid (s) fuel	Calorific value in kJ/kg
Peat (s)	13,800
Wood (dry)(s)	14,400
Brown coal (s)	16,300
Black coal (s)	23,000
Ethanol (l)	29,700
Methane (g)	39,820
Natural gas (g)	43,000
Diesel (l)	44,800
Petrol (l)	47,300
Propane (g)	101,000
Butane (g)	133,000
Hydrogen (g)	141,000

For electricity generation, if peat, lignite and black coal are abundant and close to the generator, it makes sense to use these solid fossil fuels for electricity generation, despite the low amount of energy released by peat and lignite and the lost energy from converting water in these fuels to steam. Most Western thermal power stations are in coalfields because it is far cheaper to send electricity down the grid rather than to transport coal. Electricity is converted to high voltage for long distance transport and, although losses can be up to 30%, they are far lower than if electricity was transmitted at a lower voltage. By contrast, most East Asian power stations are near ports because of thermal coal imports by cost-effective shipping from Australia, Indonesia and South Africa.

Ethanol

To put an ethanol (29,700 kJ/kg) additive in petrol (47,800 kJ/kg) is not very sensible in terms of energy efficiency, preservation of the environment or feeding people. The energy released by ethanol is far lower than the 29,700 kJ/kg as we need to subtract the energy used to make the ethanol. In the fuzzy science of economics, one law stands out like a lighthouse in a fog. The law of supply and demand.

When the demand for any commodity is increased by subsidies, tax breaks, or by mandating its use, the price will settle higher than it would otherwise have been. When ethanol (C_2H_5OH) is subsidised or mandated, its production becomes more profitable and demand for ingredients for ethanol manufacture increases. Consequently, their prices will also rise and fewer ingredients will be left for animal or human consumption. Thus more cereals, corn, sugar cane, sugar beet, soybeans and palm oil will be diverted from food production to ethanol production. As a result of higher prices for ethanol ingredients, more marginal land will be deforested and cropped for bio fuels and the production costs will be higher.

The end result is that more diesel fuel will be used for tractors and trucks to produce more crops and the produced ethanol will be mandated as an inferior car fuel. A huge amount of energy is used to transport crops while crushing, fermentation and distillation also use energy. Balancing the energy books show that there is a deficit of energy for ethanol production. The same can be shown for methanol production.

The production and burning of ethanol has resulted in the increased clearing of more land for food production. This is not really very environmentally friendly. It is well known that clearing large areas of land changes local weather and creates extinctions of locally endemic species. President Carter once advocated that America grow its own fuel supply. Furthermore, production and the burning of ethanol does absolutely nothing for the global climate as the total carbon dioxide emissions in crop growing, transport, crushing, fermentation and distillation exceeds the savings from petrol burning. Fermentation of sugars to produce ethanol produces carbon dioxide, as does burning of ethanol.

Balancing the carbon dioxide budget shows that the production and use of ethanol in fuels actually increases carbon dioxide emissions to the atmosphere. Why would the greens want to cause food shortages and add carbon dioxide to the atmosphere so that they can feel morally virtuous? This Faustian deal only makes sense to greens and to certain denizens of parliaments. For some of us, it is a dreadful waste to burn ethanol in an internal combustion engine rather than in a human body.

Methane

It is little wonder that the most abundant hydrocarbon gas (methane CH_4, 39,820 kJ/kg) is leading an energy revolution in the US. Gas prices have more than halved and the stage is set for another radical transformation in lifestyles over the coming decades. In the post-OPEC resource-rich world we are entering, we should enter a new post-alarmist Age of Plenty as we harness energy for our employment, convenience, comfort and quality of life. Goodbye pessimism, Earth Hour and "peak oil". However, the greens are doing everything to stop methane gas exploration and production. One can only speculate why the greens try to stop people growing out of poverty, having a better quality of life and having more employment.

Methane is trapped in coal, limestone and deep shale-siltstone-sandstone sequences and is produced at surface from the decomposition of vegetable matter and dung. Methane leaks out of rocks, soils, swamps and the tundra. A trace amount of methane leaks from industrial petroleum and gas plants, most of the waste industrial methane is burned.

A huge future source of methane is methane hydrate, frozen water-bearing methane that occurs in shallow marine settings. In polar areas, methane hydrate occurs in sediments that are at 0°C or less whereas at lower latitudes, it occurs in sediments from 300 to 2,000 metres depth where the bottom water temperature is less than 2°C. What this means is methane hydrates are present on almost all continental shelves and shallow marine basins. Methane hydrate is a huge long-term energy source and the Japanese government agency JOGMEC had already drilled experimental wells for methane hydrate exploitation.

The US Information Administration estimates that methane hydrates contain more energy than all other fossil fuels combined. They could hold between 10,000 and 100,000 trillion cubic feet of gas. By contrast, they estimated that there are 7,000 trillion feet of recoverable shale gas. Some of us have no concerns about "peak oil" or "peak gas". "Peak oil" was calculated on the basis of successful vertical oil wells. Many old and exhausted oil fields are now being drilled with horizontal wells and fracking of these wells has produced huge new reserves.

A popular doomsday scenario has been that we have reached "peak oil" and that crude oil production will now significantly decrease. Those supporting "peak oil" did not consider horizontal drilling in old oil fields, fracking, coal seam gas and methane hydrates. As a result of horizontal drilling, the US has moved from being dependent upon oil imports to self-sufficient. The limiting factor for cheap efficient energy on planet Earth is common sense.

Methane fuel is cheap, abundant, easy to handle and has a big bang for its buck. Furthermore, if you think that carbon dioxide is a danger to planet Earth then, during burning, methane has only one carbon atom to oxidise to carbon dioxide and the hydrogen oxidises to water. This explains why the shale gas revolution in the US has decreased their carbon dioxide emissions. The US Congress (but not President Obama) got smart and did not spend years and millions of dollars agonising over bankrupting treaties to reduce carbon dioxide emissions. The free market drove energy efficiency, reduced costs and reduced carbon dioxide emissions. Not treaties. Not legislation. Just common sense, engineering, innovation, personal freedoms, free markets and competition.

Oil and gas

Natural gas is abundant but needs far higher pressures to store and transport than methane. This costs energy and money. However, if propane (C_3H_8) and butane (C_4H_{10}) can easily be separated from shale gas and natural gas (i.e conventional oil field natural gas), then they are highly efficient cheap combustible fuels. Propane and butane are also produced as a by-product from the cracking of crude oil. This is why cylinders of compressed propane and butane are a highly efficient method of distributing and providing energy for domestic cooking, hot water and home heating. They are especially efficient for providing energy in remote locations.

Furthermore, because of the high proportion of hydrogen atoms to carbon atoms, these fuels burn to produce large quantities of water vapour and a small amount of carbon dioxide. Natural gas contains variable quantities of water and carbon dioxide (from less than 1% to over 80%). The water is separated and recycled, the carbon dioxide is separated, vented to the air and naturally recycled in plants. The costs of separation, compression and transmission of gas consumes energy and effectively lowers the energy released from burning of gas.

Burning of diesel and petrol produces large amounts of energy. This is why they are used for most means of transport. Both diesel and petrol internal combustion engines are more efficient than decades ago (e.g. by using electronically-timed fuel injection). As a result, diesel engines emit fewer sooty particulates and petrol engines emit less unburned petrol and nitrogen compounds. For example, some 50 years ago a stationary car emitted more evaporated hydrocarbons in an hour than modern cars do now driven at high speed for an hour.

The cost of exploration and drilling oil and gas wells is horrendous. A single vertical onshore well can cost more than $20 million and an offshore well can cost $100 million. Only one in ten wells is successful. After such a large amount of capital is invested, any discovered crude oil needs to be separated from water and gas. One of the most common gases in crude oil is carbon dioxide. Other gases such as rotten egg gas (H_2S), nitrogen, hydrogen and helium sometimes occur with oil and natural gas. Crude oil then is transported by truck, ship or pipeline to a refinery and cracked using catalysts and energy into various fractions such as tar, heavy fuel oil, diesel, kerosene, petrol and gas. Each of these fractions has a downstream use and needs to be transported or further refined. Sweet crude oil contains little sulphur and commands a premium.

Crude oil refineries cost billions, the only efficient ones now are the super refineries. Many countries have closed their smaller inefficient refineries and are now exposed to risk in an international conflagration because of dependency on sea trade for refined products. For example, if one more refinery in Australia closes, then military planes, tanks and heavy machinery will be dependent upon imported refined fuel. If supply lines are blocked, it just does not matter how good the military machine might be, it cannot have any effect without fuel. Large volumes of refined products need to be imported constantly over long periods of time to keep the wheels of domestic life and industry spinning.

It is no wonder that the French reduced their exposure to crude oil in the 1970s due to the Middle East oil crises (1974 and 1979) and built nuclear reactors that now provide 80% of France's electrical energy. Costs and exposure to the source of most of the world's crude oil, the unstable Middle East, is one of the reasons why the US has embraced shale gas.

Exposure to risk is one of the reasons why many places, especially the UK and European countries, should be large shale gas producers. If the Middle East or the former Russian Federation blow up, no amount of puffing by the wind for electricity generation will help these exposed countries because most of them have only have a few weeks of storage of refined products.

Hydrogen

Hydrogen gives the highest energy yield per unit weight. No wonder there has been an attraction to hydrogen as a transport fuel. However, it is totally energy inefficient and not practical. Elemental hydrogen is a very rare gas on Earth, traces leak out from rocks and volcanoes and most of the accessible surface hydrogen is in water. To produce and separate hydrogen from oxygen in water requires a huge amount of energy, this hydrogen needs more energy for compression and, if we look at the energy costs, these factors greatly reduce the energy yield per unit weight of hydrogen. Hydrogen is a very difficult material to handle and the safety risks are high.

Base and peak load electricity

Base load electricity means the minimum amount of power that must be reliably and continuously available to users, be they domestic, commercial, industrial, government, service, agricultural or manufacturers of stainless steel teaspoons. Base load varies throughout the day and the year but the variations can usually be predicted with some accuracy.

For economic and technical reasons, base load power generators are normally run at or near full capacity at all times. To ensure protection against unexpected failures, a spinning reserve of generators are run at full speed but unloaded. This is needed because it takes time to bring a generator up to speed and load and in a modern economy an unreliable power supply that causes major economic disruption cannot be tolerated. Most base load generators are driven by burning coal to produce heat that is used to produce steam to drive turbines. The rest of the base load power is from nuclear, gas, geothermal or hydroelectric generators because these are the most economic, reliable and tried-and-proven methods available. Ideological power such as wind, solar, wave or tidal does not provide the full base load power required anywhere in the world.

In addition, to cover periodic or unexpected spikes in demand, peak load generators are used. As they are not in continuous operation, they normally are gas turbine, liquid fuel or hydro generators that are faster to come up to load. They are also more expensive to run.

The best transport and industrial fuels are ethanol-free petrol, diesel and gas for road transport and methane and natural gas for domestic and industrial use. This energy mix is pretty well what we have now because these fuels are abundant, safe, cheap and energy efficient. This is market driven, not driven by some green ideology that is unreliable and subsidised.

If "renewables" had a high energy density, were cheap, unsubsidised and reliable, then in any modern industrialised country they would be competitive and would have replaced conventional energy sources. They have not. Forget electric- and hydrogen-powered vehicles, the costs of making the electricity and hydrogen are prohibitively expensive, the fuel range is small and the vehicles need to be off the road for too long for recharging. They failed decades ago and the basics have not changed.

The energy for the smelting of metals for your stainless steel teaspoon is provided by base load power. For the smelters, the fuels that can convert oxides to metals are peat (inefficient), wood and charcoal (tried before, destroyed Europe's forests, inefficient), brown coal (inefficient), black coking coal (efficient, abundant, cheap, universally used), ethanol (a terrible bloody waste), methane and natural gas (efficient), diesel (for specialist metal smelting), petrol, propane, butane and hydrogen (far too expensive).

Materials for your stainless steel teaspoon are mined, smelted and manufactured by the most efficient and cost-effective methods known, these methods have developed over thousands of years of experiments, practice and knowledge. If the cost structures change, so will the technology. If greens can devise a cheaper and more efficient way of making a stainless steel teaspoon then I invite them to go and make their millions.

Electricity generation

The logical conclusion is that for electricity generation, we need a mix of the cheapest tried and proven methods using energy density, energy yield, energy availability, reliability, flexibility and energy security. A mix of nuclear, coal, gas, hydro and geothermal electricity is ideal for base load power whereas gas, hydro and geothermal are the best for peak load power. Relying on just one electricity generating method has risks. For example, during droughts, water for hydro electricity is greatly reduced as is water for cooling towers of thermal coal-fired power stations.

There are also supply risks with gas. At times of political tension (e.g. 1996, 1999, 2014), Russia has closed its pipelines to the Belarus, Ukraine, Poland and Europe. *Forbes* magazine claimed that the 2014 trouble between the Ukraine and Russia was due to gas. It is in Russia's interest to keep Ukraine and Europe hooked on Russian gas at prices just low enough to quash incentives to drill and frack for shale gas. Russia's state-run news and propaganda outlets have for years disseminated articles critical of fracking and supported opponents of the technique.

Maybe President Putin has taken the Crimea as a kind of hostage, collateral to hold against what Ukraine owes Russia for gas. The desperation of Putin's actions underscores the threat that shale gas development really does pose to Russia's gas-fuelled diplomacy. Russia may no longer be a military superpower but it is certainly an energy superpower and energy can be used to achieve the same ends. Europe is exposed to a potential gas shortage, gas is the backbone of the chemicals industry and Europe should immediately embrace shale gas and abandon green ideology promoted by the Russians.

An EU shale gas revolution has not been stopped by armies of chanting greens. It has been stopped by greens in suits in Berlin, Paris, Brussels and London demanding that shale gas drillers jump through gold-plated regulatory hoops. Their aim has been to delay fracking such that it dies a slow commercial death. Such impediments have not been placed in the way of subsidised wind, solar and biomass burning schemes. However, thanks to President Putin's action in the Ukraine, the EU and UK are now rethinking their positions on energy independence and local shale gas. There is a good chance that President Putin has kicked an own goal and done the EU a favour.

The majority owner of the chemicals giant Ineos has written to the European Commission president warning that the chemicals industry is heading for the same fate as the defunct European textiles industry unless energy costs inflated by green taxes are reduced. He claims that six million jobs could disappear in Europe in the next decade as Europe's factories rapidly close or move to the Middle East and the US. A Russian gas crisis would only accelerate the job losses.

Alternatives?

Fossil fuels for smelting produce energy and reduction by burning carbon compounds. Carbon is essential to make your stainless steel teaspoon and one wonders how the greens' decarbonised world will actually operate. Those who advocate a low-carbon future as their latest unreasoned fad are very happy to claim that fossil fuels cause pollution, environmental damage and climate change. They point out that the Sun provides 175,000 terrawatts of energy, geothermal 40 to 50 terrawatts and gravity 3 to 4 terrawatts. That may be correct but they never comment on how much of the Sun's energy can actually be converted to electricity, how much land area that solar generators sterilise, the environmental impacts of solar power and the crippling costs of solar energy. This I show later.

Geothermal power can be noisy and release smells (e.g. rotten egg gas) and deadly toxins such as mercury, selenium, tellurium and thallium. Some countries have a contribution of geothermal power to the base load power generation (e.g. Iceland, New Zealand, Philippines, Japan, Italy, Mexico). Most countries don't have hot volcanic rocks or a high geothermal gradient for the generation of geothermal power. The environmental costs are very high. Only selected parts of the world can use tidal or wave power, a very low-density power source for electricity generation. Most forms of "renewable" energy promoted by greens are low density and massive high capital cost infrastructure is required over large areas. If a low-carbon future is your thing, then there is really only one form of reliable high-density long-term environmentally friendly form of energy. Electricity generation from nuclear fission of uranium.

If the greens want to tread another path in this highly competitive world, then they are quite welcome to spend decades gaining an education, and doing research and development to devise another energy system and alternative smelting systems for the making of your stainless steel teaspoon. Don't wait up. And don't use a wooden spoon instead. The chances of a deadly bacterium hiding in the pores of the wood are high. Then again, if you are a true green who leads by example, then please use a wooden spoon and save water by not washing it.

Kicking an own goal

No matter how much electricity is generously subsidised by the poor suffering taxpayer, the lack of grunt and unreliable electricity from "renewable" energy schemes is such that smelters could not operate. At present, there is no way large amounts of electricity can be stored for when it is needed. Smelters for your stainless steel teaspoon's metals need large amounts of energy 24/7. This energy can only come from coal, gas, nuclear, geothermal and hydropower. Constantly. For 365 days of the year. Year-in year-out for decades. No smelter can operate intermittently waiting for a puff of wind, a bovine methane-rich fart or when the Sun decides to come out from behind a cloud.

If we are to eat with a stainless steel teaspoon, then we need conventional power supplies that consistently generate large volume electricity for a very long time. It takes days to fire up a smelter and they just cannot be turned on when the wind blows and turned off when there is no air movement. If smelters are to be reconfigured, then thousands of years of practical and theoretical knowledge must be supplanted in order to create a new process to convert rock into metal.

If greens really wish to have "renewable" energy, they should lead by example. They would need to work hard to end up in the top 5% of school leavers, gain entry to study engineering at university, after a first degree they would have to undertake a research degree and then spend decades competing with engineers all over the world using their knowledge to try to make the world a better place with cheap efficient electricity and environment-saving inventions. They don't. They sit on the sidelines, complain and offer no tried-and-proven new energy generating systems derived from decades of their own hard work. They take all the benefits of the modern world that they are knowingly trying to destroy. As has been shown time and time again, the greens' solutions to perceived problems damage the environment. One of the best examples is electricity generation by wind.

Cost of conscience

The big future election issue in the Western world is the cost of energy and its impact on unemployment. Hare-brained unproven green schemes using "renewables" have added to home energy, transport and food costs. Political coalitions between centre left and green parties have raised the cost of energy. Unnecessarily inflated energy costs have now pushed up the punter's living expenses so high that politicians are starting to respond. A good politician is a frightened politician.

High energy costs have driven energy-intensive manufacturing and smelting to low-cost energy countries resulting in job losses. A good example is the moving of BASF from expensive Germany to the US as a result of the low-cost energy from US shale gas. Steel (e.g. UK), aluminium (e.g. Australia) and motor vehicle manufacture (e.g. Australia) have closed, mainly because of high energy costs. Airlines pay billions each year for fuel: carbon taxes and high energy costs have severely affected the bottom line.

Kyoto capers

The Kyoto Protocol was adopted in 1997 and was in force from 2005 to 2012 as the climate yardstick. It was so important that the future of humanity depended upon the Protocol. So important that it disappeared without even a whimper on the last day of 2012. The Protocol was to lock signatories into reducing their greenhouse gas levels relative to their 1990 emissions.

But, when ideology and reality clash, reality always wins. Japan signed up for a 6% reduction in greenhouse emissions yet saw a rise of 7.4% during a period of economic stagnation and an increase in nuclear power generation. Australia experienced a vigorous growth, it signed up for an emissions increase of no more than 8% yet its emissions over the two decades increased by 47.5%. Both Canada and Australia were enthusiastic backers of the Kyoto Protocol. Canada signed up for a 6% cut and managed a 24% increase in 1990-2010.

The EU met its emissions targets due to economic stagnation, the closure of inefficient Soviet era industries and a carbon cap-and-trade scheme that allows industry to move production abroad and collect payments for "carbon credits" from the poor European taxpayer.

The US saw economic and population growth between 1990 and 2010 yet only had an emissions decrease of 10.3%, mainly due to the shale gas revolution using technology that the EU and many other countries refuse to embrace.

Canada led the world and withdrew from the Kyoto Protocol as did New Zealand, Russia and Japan. The world's largest emitters, China and USA, were not silly enough to sign up in the first place. Now only Australia and the EU remain members of this exclusive Kyoto Protocol club that distorts world energy markets, is scorchingly expensive and promotes legislation that destroys jobs. Like other great all embracing global environmental programs led by the finger-wagging bleeding heart armchair greens, costly job-killing failure is guaranteed.

There have been 18 Kyoto gab fests in very pleasant parts of the world, hundreds of delegates have been sent on jaunts by signatories, these have cost about $1 billion and each gab fest has made a momentous decision. They decided to … meet again. And the taxpayer pays for this perverse joke.

UN hypocrisy

A recent UN report admonished the world for not becoming "climate neutral" fast enough to avoid the "catastrophic consequences" of climate change. I have no idea what "climate neutral" means. Does it mean that the planet is static, as argued by creationists? Does it mean that we humans actually stop natural climate changes? What catastrophic consequences? Are more plant food and the greening of the planet catastrophic? Such a statement implies that all climate change is of human origin showing that the UN is removed from science and is operating as a partisan political lobby group.

The UN's IPCC are now telling us that if we don't curb our use of oil, gas, coal and meat, then the atmospheric carbon dioxide content will rise, the planet will warm and food production will decrease. Who are these IPCC people? Are they just academics and modellers? They should have spoken to practical people. Horticulturists pump warm carbon dioxide into glasshouses. Plants grow faster, bigger and become more drought- and heat-tolerant. If these IPCC folk actually got outdoors, they would realise that plants grow more quickly in summer when it's warm than in winter when it's cooler. Thousands of years of farming and science tell us that food grows better when it is warmer. It is the same IPCC that is advocating for a meat-free Monday. What next? A food-free Friday!

This is really a hypocritical comment from the UN considering that they have failed to reduce their own carbon dioxide emissions. Each year, every human on Earth generates an average of 4.637 tonnes of carbon dioxide emissions, including exhaled air. Because of travel and energy use, a UN worker emits over 8.2 tonnes of carbon dioxide *per annum*. There are over 215,000 workers at 54 UN agencies scattered in 530 duty stations around the globe. In 2008, these UN workers emitted 1.741 million tonnes of carbon dioxide. Of this, air travel contributed to 4.02 tonnes of carbon dioxide emissions per worker. This was 48% of the total. Two years later, carbon dioxide emissions had risen to 1.766 million tonnes with air travel accounting for 51% of emissions.

Until the UN drastically reduces its own carbon dioxide emissions, we should treat every scare story they promote as just hot air.

Own goals

Globally, in 2012 solar and wind power were subsidised to the tune of $60 billion. And the climate did not change. For all this extra money spent on inefficient energy, only 0.3% of global energy was from wind and 0.04% from solar. The carbon dioxide emissions savings that translate into climate benefits were just over $1 billion. For every dollar invested, 97 cents were wasted. Today's "renewable" technologies are expensive, are subsidised, cannot store electricity and there is no breakthrough on the horizon. This is a global own goal by greens that comes at a horrendous cost to taxpayers.

The UK and European countries have kicked some spectacular own goals. For example, EU member states will not be allowed to continue with coal- and oil-based electricity generation under one set of rules and, under another set of rules, are required to use more subsidised expensive "renewable" energy. Furthermore, as a result of green energy schemes in the EU, there has been a loss of four million jobs in Europe in the manufacturing sector. Well done greens.

Europe's carbon market has collapsed. The scandal prone $150 billion market has taken less than a decade to end up an unruly monster. Far too many permits were produced, the market is in oversupply, the price collapsed and, for some odd reason, traders seemed more interested in a profit than saving the planet. The Labor government's carbon comrades in Canberra tied their price in carbon to the European price which promptly fell through the floor. The scheme could only have worked if China, India and the US signed up. They didn't. The carbon market could only have been thought up by ideological fools totally out of contact with reality.

This EU directive has fallen apart because of electricity shortages and the weather. In December 2013, Germany's wind and solar power generation came to a standstill. More than 23,000 wind turbines stood still. One million photovoltaic solar systems went on strike. For one whole week, Germany had to depend on nuclear (imported from France), coal and gas electricity generation.

German citizens paid a total of €20 billion in 2013 to promote "renewable" energy. In 2014, the figure will be nearly €24 billion This money passes through a large number of unproductive sticky fingers known as administration before actually building an inefficient structure. German social campaigners state that some 800,000 households can no longer afford to pay their electricity bills. This number has risen dramatically in recent years.

In Germany, household electricity prices have risen by 80% since 2000 and some seven million households now live in energy poverty.

The German energy crisis was man-made. By Germany and the EU. It is ironic that the EU, whose energy policy is largely based on the promotion of "renewables" and a target cut in emissions by 20% by 2020, has ratified the Kyoto Protocol yet has not been able to reduce carbon dioxide emissions. The US did not sign the Kyoto Protocol and has reduced carbon dioxide emissions.

In the UK, the green goals for "renewable" energy have produced rising energy costs and now 17% of UK households are energy poor. At times, the greens claim that they are concerned about poverty. Are they? Their actions push people into poverty. The UK greens proudly announce that households have decreased electricity use by 10% since 2005. They just happen to forget to mention that there has been a 50% increase in electricity prices to pay for increasing the share of "renewables" from 1.8 to 4.6%. This price increase hits the poorest hardest, as with green taxes, because electricity is essential and now takes up a greater proportion of a small budget. The poor have reduced their energy consumption, not the wealthy.

In the UK over the last five years, home heating costs have risen 63%, real wages have decreased and an increasing number of the poor spend more than 10% of their income on energy. Energy poor pensioners are spending their days riding in heated buses to keep warm, a third are leaving parts of their homes cold and rugging up with hats and scarves and blankets inside their homes and they are forced to stay in bed longer because of the cost of energy. Is this is green policy helping the poor or the triumph of ideology over tried-and-proven systems?

The use of Russian gas is the only short-term solution to the intermittent electricity generated from wind and solar. This creates a new energy risk. Germany's security is already compromised by dependence upon Russian gas. If Germany continues to shut nuclear and coal generators to become fully dependent upon "renewable" energy, dependence on Russia will increase further. The EU now is scrambling to find ways to be less dependent upon Russian gas.

Greenpeace and other unelected activist groups have forced the shutdown of nuclear power in Germany and have stopped fracking for gas. What's left? Good old coal. King Coal. Over the next two years Germany will be building ten new unsubsidised coal thermal power stations. Europe is enjoying a coal boom with the building of new power plants and mines in Germany, Czech Republic and Poland. This goes against the EU rules for limiting carbon dioxide emissions and for "cleaner" energy.

Coal-fired power stations in Germany are replacing the eight nuclear power stations that were shut down because of green pressure and, because Germany has a huge solar and wind generating industry, the unreliability of these ideological power sources is such that Germany has now increased its carbon dioxide emissions by building new thermal coal power stations. Another own goal.

However, German electricity prices are now almost twice those of the US and it is hurting. The coal boom in Germany is a result of Greenpeace's political success. Another own goal, this time for Greenpeace. Even the co-founder of Greenpeace, Dr Patrick Moore, has washed his hands of them because he claims they are now no longer an environmental group and are a thuggish socialist political pressure group.

However, Greenpeace provides constant entertainment. The Greenpeace protests against the culling of man-eating sharks in Western Australia were conducted on the beaches. If they loved these sharks so much, then why were they not neck deep in the Indian Ocean shouting and waving their banners? I'm sure a shark would never think of devouring someone that loved them so much.

Denmark had been a very enthusiastic supporter of wind energy. In 2004, Denmark decided to build no more wind farms because it was producing the most expensive energy in Europe. Denmark could see the financial writing on the wall. Although the Danes had become dependent upon wind energy, they found that when the wind did not blow they could not buy wind-generated electricity from north Germany because the weather conditions were the same. They resorted to buying more reliable hydro- and nuclear-generated electricity from Norway or nuclear-generated electricity from France at highly inflated prices. What a wonderful opportunity for the Norwegians and French to skin the Danes alive. And they did. When the wind was strong, the power could not be sold because it was also strong in north Germany. This electricity had to be given away. Denmark now has green taxes that account for more than 50% of an electricity bill. Another own goal.

The US Environmental Protection Agency has engaged in a campaign essentially to regulate coal-fired electricity generation out of existence in the USA. Twenty-nine US states and the District of Columbia now have "renewable" energy mandates and many are trying to impose cap-and-trade programs. If indeed humans are changing climate, funds that could be dedicated to helping people prepare for and adapt to climate change and extreme weather events are wasted on futile attempts to stop what might (or might not) possibly happen in 50 or 100 years time. The US alone spends $7 billion each year on "warming studies" which, in truth, is nothing but a huge money laundering operation as no real science is conducted. Vapid alarmist reports are the only product generated.

Africa is a green utopia. They already generate 50% of their energy from "renewables" such as twigs and dung. However, no green wants to join the three billion people worldwide that rely on burning twigs and dung for heating and cooking. The resultant indoor pollution is the biggest environmental problem on Earth and kills 4.3 million people annually. The Centre for Global Development suggested that for an investment of $10 billion, 20 million African could be lifted from poverty by using "renewables". However, if $10 billion were spent on gas generation of electricity, 90 million Africans would be lifted from poverty. Is this the green agenda? To leave 70 million people in darkness and poverty.

In 1971, China derived 40% of its energy from similar "renewables".

The growth of China by using coal has lifted 680 million people out of poverty and only 0.23% of China's energy is from tokenistic wind and solar.

There is a bottom line. Green schemes result in red ink, poverty and death.

China and India

In terms of carbon dioxide emissions, it does not matter whether the EU bans or uses coal for electricity generation. China continues to increase its carbon dioxide emissions from coal. During 2013, China added 100 million tonnes of coal production capacity. This was six times more than for 2011 and is about 10% of the US annual use of coal. China is closing small polluting mines, factories and power stations that add choking sulphurous and particulate pollution in the atmosphere. China is restricting imports of high-ash high-sulphur brown coals and is concentrating on mega mines and coal-fired power stations. Over a five-year period from 2011 to 2015, China plans to add an additional 860 million tonnes *per annum* of coal capacity. This coal is burned in smelters and for electricity generation. It is predicted that Indian coal-fired electricity generation may eventually overtake that of China. It is coal that brought the Western world out of poverty and it is immoral of the greens to attempt to stop the same transition to prosperity occurring in China, East Asia and India.

Coal versus ideology

King coal cannot be killed off by ideology. There is a bottom line. Coal is the biggest source of fuel for generating electricity in the world except France, which has nearly 80% of electricity from nuclear fission, and Saudi Arabia which burns oil for electricity generation. Global coal exports are growing quickly. This demand is being stoked by the rise of electricity-hungry middle classes in emerging economies in East Asia,

China and India. At the current rate of growth, by 2020 coal is expected to surpass oil as the dominant fuel source. Without coal we cannot support modern society.

The coal industry is highly internationally competitive. In most parts of the world it is not subsidised. Ideological energy is subsidised and many of the subsidies are actually borrowings. There is a limit to borrowing. There is a limit to spending more than is earned. Eventually, the well runs dry. It has started to run dry in the UK and Europe.

Wind power

Electricity from the wind is totally unreliable, uneconomic and degrades the environment. Wind energy neither decreases carbon dioxide emissions nor changes global climate. No wind farm could operate without generous taxpayer subsidies and increased electricity charges to consumers and employers. These subsidies are given irrespective of whether the wind farm produces any consumable energy or not and are paid even when a wind farm is shut down due to strong winds. Wind farmers have been more successful in harvesting massive subsidies from taxpayers than harvesting the wind. The subsidies in Australia are paid per megawatt generated via a "renewable" energy certificate. More bureaucratic jobs are needed.

No green act of faith can control nature and make the wind blow when and where energy is needed. We have wind farms because unelected and thus unaccountable green political pressure groups claim that wind power is "renewable", is environmentally friendly, does not emit carbon dioxide and is good for the environment.

All these claims regarding wind energy are demonstrably wrong. Noise generated by an organised minority has resulted in a disorganised majority suffering and paying as a result of the minority's successes. Wind farms are permanent memorials to a period of collective madness that overtook the Western world when politicians responded to pressure from unelected, unaccountable, noisy green minorities. Plato said: "The penalty good men pay for indifference to public affairs is to be ruled by evil men."

Wind farms produce less than 30% of their nameplate capacity, often at times of low electricity demand and low electricity prices. No carbon dioxide-emitting coal-fired thermal power station has been replaced by a wind farm. Not one! And why? Reliable tried and proven low-cost efficient electricity generation from coal is needed as backup because most of the time the wind does not blow or it blows too strongly. Coal-fired power stations take 24 hours to fire up and they just can't be turned off and on depending upon whether the wind decides to blow or not. In still cold weather, wind farms consume electricity from coal-fired power stations to stop lubricants freezing.

Wind power is subsidised in order to artificially compete with cheaper coal-fired electricity, made more expensive by green legislation. There is no shortage of energy, only a shortage of common sense that results in a shortage of cheap electricity.

Industrial economies and urban areas need low-cost efficient electricity to function. Eventually, subsidies will run out and the countryside that was once beautiful will be left with defunct wind farms as a memorial to arrogant green stupidity. In many places, there is no bond held for decommissioning wind farms and land rehabilitation.

All mining operations in Western countries have a bond for environmental rehabilitation after mining, in case the mining company goes broke. Not so for wind farm or solar companies in many jurisdictions. As soon as subsidies stop, these companies will go broke. Many already have. Defunct wind farms already pollute the Californian countryside from failed wind schemes that sent wind farm companies broke. There was no legal requirement to remove the infrastructure in some jurisdictions. Is this environmentalism?

Wind farms have the lives of parasites. They cannot produce

continuous electricity without coal, gas, nuclear, hydro or geothermal backup. They freeload by attaching themselves to an existing electricity grid built and paid for by those using conventional energy. They seldom contribute to the maintenance costs of the transmission network and consumers are forced to pay a feed-in price for their unreliable output. These green energy parasites are still paid during those many times when they produce no power and drain electricity from conventional sources when it is too cold.

Each January-February, the Northern Hemisphere has a cold snap and the wind just does not blow. People die. In southeastern Australia in January 2014, the grid needed 12,000 megawatts at peak when the temperature was more than 40°C for days. The 28 wind farms in southeastern Australia could only provide 128 of the 12,000 megawatts required and it was coal that provided the electricity for air conditioning. When wind farms were needed to provide much needed electricity for cooling, they only operated at less than 5% capacity.

Furthermore, during the 45°C heatwave on 14 January 2014, South Australian electricity wholesale prices spiked at $10,515 per megawatt hour. The power grid collapsed and many people were without electricity. This was by far the world's most expensive electricity and is just a tiny little bit above the wholesale long-term spot price of $70 per megawatt hour. In South Australia, 40% of the electricity is supposed to come from wind power. It doesn't.

If the wind were constantly blowing at 11 metres per second at every wind farm in South Australia spread over hundreds of kilometres, then the nameplate capacity of 1,203 megawatts would be generated. This does not happen. Greens state that the wind is always blowing somewhere over such an extensive area so power is always being produced. Reality is different and this does not happen.

In reality, only 60% of South Australia's notional generating capacity is available to service demand when wind watts go walkabout over 100 times a year. When there is no wind, open cycle gas turbines (at $300 per megawatt hour) and 65 megawatts of diesel generators at the defunct Adelaide Desalination Plant kick in to generate electricity and make a killing at the expense of the consumer. No wonder South Australia cannot attract industry investment.

When more power is needed in extreme hot or cold periods, the answer is certainly not blowing in the wind. Wind farms just take a holiday. Incidentally, it now appears that the December-January hot periods in Australia are described as "extreme weather". When I was a child, it was called summer.

Wind turbines all tend to produce peak power at the same time when winds are strong. They also all produce nothing when there is no wind. This surging creates huge transmission network problems and, at times, the network is over-capacity. Because of this, wind power is very sensitive to wind speed and can only operate at low wind speeds and therefore is the lowest quality power for the grid. At other times, it is under-capacity.

Wind power generation schemes were designed by Enron and embraced by green politicians. Power engineers would not have designed such an inefficient system. Politicians have created a business opportunity for wind farm owners to skin the punter alive and they do. Who can blame them? We can blame politicians and election time is when we can make those responsible accountable. The easiest and most environmental friendly way to kill these parasites is to stop subsidies.

Energy poverty

This warm embrace of feel-good highly expensive "renewable" wind energy has left the most vulnerable citizens out in the cold. Literally. In Germany, charities report the power is cut off from more than 300,000 households each year because consumers can't afford to pay the high costs of "renewable" green electricity. Some 800,000 Germans now have energy poverty. Electricity has now become a luxury item for many. German consumers now will be forced to pay annually more than €24 billion to subsidise electricity from solar, wind and bio fuel generating plants that produced electricity at a market price of just over €3 billion. Because of the green dream, Germans now have the highest electricity prices in Europe.

In the UK, green levies for "renewable" energy are causing energy poverty for 2.4 million British households. In the UK, there are some 6,000 wind turbines with about 1,000 offshore. In the 2012-2013 UK winter, there were 35,000 additional deaths, this correlates with the increase in wind turbines and it correlates with the increasing number of the people facing energy poverty. This translates as six sick, elderly or vulnerable people killed each year for every wind turbine or six deaths per megawatt of wind power generated. Every winter, there are more people who die from hypothermia in Scotland than in Finland, a far colder part of the world. I am sure that those greens who can afford the increased costs of electricity get a warm inner glow from such statistics. The sick, elderly and vulnerable that suffer energy poverty actually subsidise their own deaths, thanks to the greens.

In the 2011-2012 winter, tens of thousands of trees disappeared from parks and woodlands across Greece. Impoverished residents did not have money to pay for electricity and turned to fireplaces and wood stoves for cooking and heat. The same has occurred in Germany. The combination of a cold winter and rising energy costs has forced people to go collecting wood in the forests for home heating and cooking.

Is this the future? Europeans huddled around environment-destroying wood stoves to keep warm because electricity is too expensive and subsidies are paid to the wealthy who own wind farms and solar panels. This is the way Europeans lived centuries ago and yet the greens claim that they are "progressive".

What about the Sun?

Climate is changing and it is related to that great ball of energy in the sky called the Sun. The Sun is a giant nuclear fusion reactor and all life on Earth relies on that nuclear power. The Sun could be on the threshold of another Maunder (1745-1715) or Dalton (1790-1830) Minimum event as Solar Cycle 24 is the weakest for 50 years. It was in these solar minima that the Earth was very cold, major rivers froze and crop failures across the Northern Hemisphere led to starvation and death. If this solar event occurs again, Europe will need more than wood for warmth and cooking. The wealthy will survive, as they did in past solar minima. This is contrary to the greens' avowed policy of stopping poverty.

In Australia, the "renewable" energy share of total energy has remained at 6% for two decades. During this time, carbon dioxide emissions from human activity have increased by 58%. Most of the "renewable" energy is from older hydroelectric schemes not subsidised by the "renewable" energy scheme. Greens would not allow hydroelectric schemes to be built today even though it is "renewable" energy. In 2013, wind produced 0.4% of energy and solar 0.1%. In 2013, Australia spent more than $6 billion on "clean" energy. What really is "clean" energy?

The creation of electricity by coal-fired generators is very clean energy, the carbon dioxide and water vapour released are the foundations of life on Earth and hence the whole misleading jargon of "clean" energy will cost the taxpaying community billions of dollars. Green schemes don't use money from taxation revenue, they use debt. It needs to be paid back. By the poor innocent punter's taxes. The green ideology is not "alternative" energy, it is a very expensive alternative to energy.

The small amount of wind and solar electricity developed in Australia has resulted in a disproportionate increase in costs and a decrease in efficiency. Australia was looking to join the EU emissions trading scheme (i.e. the one that did not reduce global carbon dioxide emissions). The collapse in the carbon price is the market showing how trading in carbon is inefficient and ineffective. Markets generally get it right.

Blowing in the wind

Humans have used energy from the wind for thousands of years. The best example is the use of wind for sails on boats. Four thousand year-old

Egyptian pottery shows large sails on ships and windmills from hundreds of years ago used canvas sail blades for grinding of grains and corn. Many defunct canvas-sailed windmills are preserved in Mediterranean countries. No passenger or freight ships use sails today, wind transport is far too slow, unreliable and expensive. Wind transport is dangerous in areas of high winds (e.g. Cape Horn). In isolated rural areas, windmills are used to pump small amounts of water for stock and domestic use because they are an inexpensive, low maintenance, low energy system that allows water storage in tanks. The wind industry leads the public to believe that wind turbines are just harmless windmills, like those in rural Australia. Many rural areas also use diesel and solar energy for pumping water. In isolated areas, it is only diesel pumps that have enough power for moving large volumes of water for irrigation.

There are daily variations in wind caused by solar energy heating the land, sea and air. Wind measurements show that wind directions and speed are constantly changing. In as little as a second, wind speed can double and the wind direction can reverse. The rotation of the Earth means that the amount of solar energy changes during a 24-hour day. During the day, air temperature rises more over land than over water due to the different thermal properties of soil and water. Hot air over land expands, becomes lighter and rises and the cooler heavier air from over the sea replaces the air that has risen and moves across the land as a gentle zephyr. This is the sea breeze. All cricketers know about the Fremantle Doctor, a mid afternoon sea breeze used by bowlers to move the ball in flight. During the night, the land cools more quickly than the sea. Cool air sinks and moves seaward. The warmer air above the sea rises. These breezes can extend up to 200 kilometres inland in the tropics and some 50 kilometres inland at mid latitudes.

In mountainous areas, the peaks heat earlier in the day than in the valleys. The warmer air rises from the peaks and heavier cooler air in valleys moves up to replace air that has moved upwards from the peaks. At night, the opposite happens and cooler air from mountains gravitates down valleys. In some extreme cases such as in Antarctica, air cools over upland ice and cascades down slope creating very strong perishingly cold winds that can be almost constant and up to 200 kilometres per hour. The geologist Sir Douglas Mawson struggled for years against these winds in Commonwealth Bay.

Wind farms

Large scale wind electricity generation would be needed if a modern industrial nation is to be powered by this form of "renewable" energy to smelt metals or to fabricate your stainless steel teaspoon. Wind, solar, wave and tidal are low-density forms of energy whereas uranium, oil and coal are high-density forms of energy. The argument presented by greens that a spread of wind farms produces of smoothing of the power generated, thereby providing base load power and reduced power surging. However, a study of 21 widely separated wind farms connected to the eastern Australian grid shows this is not the case and wind power is unreliable and that there are numerous times when the wind power output is zero.

According to the EU, wind farms operate with less than a 20% load factor which is less than 20% of the ideal peak or rated values of wind turbines. The Environmental Protection Agency then calculates that if a wind farm is to produce the same amount of energy as a 1,000-megawatt conventional coal-fired or nuclear electricity generating plant around the clock, the land area required at 20% load is about 844 square kilometres. A conventional 1,000-megawatt coal-fired or nuclear power station occupies a block of land of 75 to 150 acres (30 to 60 hectares) in area. Of course, it is far more environmentally friendly to clear more than 844 square kilometres of land for an ideologically correct wind farm than less than one square kilometre of land for an equivalent conventional coal- or nuclear-fired power station.

More habitats are destroyed, more wildlife is killed, more bush fires are started because of the frequent fires in wind generators, more transmission lines are needed, more roads need to be cut, more scenic areas are despoiled, more people are affected by the infra-sound and low frequency noise and more properties are devalued. Land needs to be cleared to reduce turbulence and to increase free flow of wind. Will the greens inspect this land area in its entirety on foot with the native landowners for cultural artefacts and areas of significant cultural heritage? The mining industry does such cultural surveys. Why not the wind industry?

Wind turbines convert the energy from moving air into electricity, the wind is robbed of its energy and local wind speeds, temperature and rainfall change. The more wind towers there are, the more the local weather is changed. Winds bring moist air inland, dilute and clear air pollution from coastal cities and change air temperature.

Very large turbines create low frequency microseismic ground vibrations. Geophysical mineral exploration cannot take place near wind farms thereby sterilising large areas of land that may contain resources for our future. High wind towers with long blades are an aircraft navigation risk. Current blade design causes interference with aircraft navigation and TV signals. Radar can't distinguish between rotating blades and aircraft. Efficient energy generation requires very smooth blades. In winter, ice accumulates on blades, there can be an accumulation of dead insects on the blades and as a result there can be a power reduction by up to 25% and shaking of the whole structure. This has created structural failures. Turbines fail, catch alight and create grass fires. As wind farms are a no-fly zone, aerial water bombardment cannot take place close to wind farms and the grass just burns and burns. And the greens try to tell us that wind energy is good for the environment!

If a 1,000-megawatt wind power station is to be built, there are some considerations. Can 1,000 megawatts be generated each day? Can 1,000 megawatts be generated continuously? Can 1,000 megawatts be used when it is required? Is there 1,000 megawatts of conventional electricity backup? Wind intensity and direction need to be measured over a long time period to calculate the ideal location and costs.

The theoretical perfect wind turbine can only extract a maximum of 59% of the available kinetic energy of the wind (Beitz limit). Under ideal wind conditions with high-efficiency large turbines with only a few thin blades rotating five to six times faster than the wind can only extract about 45% of the available kinetic energy. For varying wind conditions, turbines need to be placed ten blade diameters apart.

If Australia is to meet its own self-imposed "renewable" energy target of 20% using wind using wind turbines alone, the figures are daunting. The total Australian electricity production for a year is 260,000 gigawatt hours hence 20% of this is 53,000 gigawatt hours per year. With a 20% load factor, the equivalent energy production would be the same as the turbine operating at peak efficiency for 4.8 hours per day. Hence, for a 1,000-megawatt wind farm, 4,800 megawatt hours per day are generated (i.e. 1,752 gigawatt hours per year). As 53,000 gigawatt hours per year are required and a 16-gigawatt wind turbine farm produces 1,752 gigawatt hours per year, then 30.25 wind farms are needed.

Using the common Vesta turbine that generates 660 kilowatts at peak performance, 1.52 turbines would be needed to produce 1 megawatt. For 1,000-megawatts, 1,520 wind turbines are required and with a 20% load factor 7,600 wind turbines are needed for a wind farm that produces 1,000 megawatts. To produce 20% of Australia's annual power requirements under "renewable" energy targets, 229,900 wind turbines would be required. Wind turbines are spread apart because they interfere with each other and steal wind. Usually nine turbines occupy a square kilometre. To generate 20% of Australia's annual power requirements under "renewable" energy targets, 25,531 square kilometres of land would be required.

The cost of purchasing such a land area is prohibitive, to find 25,531 square kilometres of land close to high voltage grids is not possible and to clear 25,531 square kilometres of land to save the environment is the end result of the greens' "renewable" energy policy. In Australia, the wind companies have solved this little problem by paying farmers $10,000 per turbine *per annum* to lease their land. By leasing the land, liabilities are transferred from the wind farm company to the farmer.

If 100% of Australia's energy were to come from the wind, only 127,655 square kilometres would be needed! Tasmania is full of green activists and even if the whole of Tasmania was clear felled and covered with wind turbines in the name of green activist environmentalism, Tasmania's land area of 64,519 square kilometres could only provide about half of Australia's total energy needs from wind. And greens in Tasmania are only too willing to promote wind power! Have they really thought it through?

The environmental cost of wind power

Wind generating towers mainly comprise steel (about 100 tonnes for a 65 metre high tower) and concrete (260 tonnes). The rotor weighs seven tonnes. For a 1,000-megawatt wind farm comprising 7,600 turbines, 760,000 tonnes of steel and 1,976,000 tonnes of concrete would be required. Steel and concrete just don't come out of thin air and a huge amount of energy is needed to make them. For every tonne of steel 15.118 kilowatt hours of electricity is needed for manufacture. For concrete, it is 3.147 kilowatt hours. This energy resides in the steel and concrete as embedded energy. For the steel and concrete for a 1,000-megawatt wind farm, 11,490,000 and 6,218,000 megawatt hours respectively of electricity are needed. With a 1,000-megawatt wind generator, producing 1,752,000 megawatt hours per year of electricity, it would take 10.11 years to create the energy needed just to manufacture the steel and concrete. This is, of course, a minimum figure as there is embedded energy in all of the copper wiring, rare earth element magnets, light metal and carbon fibre turbine blades; in the transport costs of steel, transport costs of cement and aggregate; in the crushing costs of aggregate; in the mining of iron ore, limestone, shale and fluxes; in the costs of transporting some 100,000 truck loads of concrete; in road building and wiring; and in towers, concrete and insulation for transmission connection to the electricity grid. Massive quantities of diesel would be used for transport and construction. It really is criminal to invest all this money and effort into an inefficient problematic "solution" to a non-problem.

The correct figure for payback of just the embedded energy is probably more in the order of 15 to 20 years. Most wind farms have a rated life of 25 years and a real life of 15 to 20 years. Whatever the figure is, wind farms look like bad business. Where is the pre-feasibility study backed up with a feasibility study, as would be done for any new mine? No need for such basics because money is just lifted out of taxpayers' pockets for subsidies. In other businesses, the promoters would go to gaol.

It is interesting that green groups are quite happy to put people out of work with their ideology. When the obvious practical, engineering and financial weaknesses of wind-generated electricity are pointed out, greens then argue that wind farms will create employment. They certainly will. Wind farms create subsidised jobs that are paid for from unsubsidised workers' taxes. The net effect upon employment of subsidising some jobs at the expense of others is invariably adverse.

We all know that wind farms reduce the amount of that dreadful plant food called carbon dioxide, sinfully released by humans into the atmosphere by burning coal. But we also can calculate the amount of carbon dioxide that wind farms release into the atmosphere. The total energy required for the making of steel and concrete is 17,708,000 megawatt hours. Each megawatt hour of electricity generation from coal produces about 430 kilograms of carbon dioxide.

To make concrete, limestone needs to be heated to release carbon dioxide during the making of cement. Structural concrete contains about 14% cement and for every tonne of structural concrete, there are emissions of 180 kilograms carbon dioxide. This is mainly released during the burning of limestone to make cement. To make the cement for the 1,976,000 tonnes of concrete would release 355,680 tonnes of carbon dioxide into the atmosphere. Hence for the production of the steel and concrete for the 1,000-megawatt wind farm, a total of 7,616,440 tonnes of carbon dioxide is released into the atmosphere. And wind farms are meant to reduce carbon dioxide emissions?

Furthermore, because coal-fired thermal power stations are not replaced by wind farms thereby not reducing their carbon dioxide emissions, there is an additional 9.5 million tonnes of carbon dioxide added to the atmosphere for every 1,000-megawatt wind farm because coal-fired power stations are on standby for when the wind stops. In addition, maintenance is undertaken over a very large area of the 1,000-megawatt wind farm by diesel-powered vehicles that add carbon dioxide and particulates to the atmosphere.

Even simple calculations show that wind farms actually increase the amount of carbon dioxide that humans emit to the atmosphere. If carbon dioxide is really a pollutant, then wind farms are more polluting than many other forms of energy generation. Why do the greens claim that wind energy reduces human carbon dioxide emissions? It does not. Do greens know how steel is made? Probably not. Do greens know how concrete is made? Probably not. Are they gilding the lily? More than likely.

But wait, there's more. Just the simple matter of money. Something the greens don't worry about because it's not their hard-earned money. If a Vesta 660 generator is used, it has a maximum capacity of 660 kilowatts at a cost of about $2,000 per kilowatt installation cost (i.e. $1,320,000). With a wind farm with 7,600 Vesta 660 generators to produce 1,000 megawatts, the cost is a knockdown price of over $10 billion. This cost excludes the entire surface infrastructure such as the site preparation, concrete base, roads and transmission lines that would add a few more tens of billions of dollars.

If Australia is to produce 20% of its electrical energy from wind, then the total capital cost just for the generators is a bargain basement price of $303 billion. The figure is probably twice this because of logistics and peripherals. Where is the money for 20% of electrical energy from wind going to come from? Subsidies? Debt? Increased taxation? Electricity consumers? It certainly is not going to come from the Federal budget which is about this figure. It certainly will not come from the pockets of green activists promoting "renewable" energy.

This needs to be put into perspective. Three 225-kilowatt V8 petrol car engines or one small Cummins QST30-G2 diesel gen-set generates as much electricity as a single 65-metre high Vesta 660 wind generator. The embedded energy in the petrol and diesel engines is miniscule compared to a wind tower and generator. Furthermore, they are far cheaper to buy, cheaper to run, portable, require no grid, produce less carbon dioxide than a wind turbine and generate electricity all the time, not just when the wind blows. It is no wonder that many isolated areas use diesel gen-sets rather than wind generators.

Who pays the piper?

The UK is rather keen on destroying its beautiful scenery with wind farms. Even the Prime Minister David Cameron erected a symbolic wind generator on his home that does not generate enough energy to cook a meal. The UK has a small land area (243,610 square kilometres) and a long windy coast (12,429 kilometres or 7,723 miles). This is why coastal and offshore wind farms have been attractive (as long as they are massively subsidised). The same sort of calculations can be made for the UK showing that their land area is not big enough to generate all the required electricity from wind farms.

In the UK, wind turbine companies pay farmers up to £100,000 per turbine so there is certainly a huge amount of taxpayers' subsidy money sloshing around. In the UK, there are about 6,000 wind towers and most are visible from every high hill in the country. Less than 1,000 hectares of shale deep in one of the many shale-bearing sedimentary basins in the UK would produce as much energy as the entire UK wind industry. Fracked shale gas surface facilities would occupy less than one hectare and would provide gas for electricity generation 24/7. No inefficient subsidised unreliable wind farms would be necessary and jobs would be created in the UK where they are needed. A UK government study has shown that the lifetime cost of "renewable" energy targets in the EU is £290 billion, more than a quarter of which will come from the UK.

In Australia, each turbine at Infigen's Bodangra (NSW) wind farm produces $850,000 revenue *per annum* (including subsidies) and, in order to win over local support, $17,000 *per annum* royalty is paid into the Bodangra Community Enhancement Fund on top of the $10,000 per turbine *per annum* to the farmer.

Wild life

And what about the environmental effects of wind farms on wild life? Many UK farms are uneconomic and survive on subsidies. Others rent a few hilltops that they can't use for crops and receive generous funding from subsidised wind farm companies. To make matters even more ridiculous, wind farm operators in the UK were paid £34 million in 2011 to switch the turbines off in gales. Householders handed money over to wind power generators for doing absolutely nothing. These Soviet-style wind farm subsidies are a never-ending gravy train that mercilessly raises the cost of electricity.

Rare birds and bats are sliced and diced by the rotating blades of wind turbines. In Spain, the wildlife conservation group SEO Birdlife states that each year between six and 18 million birds are killed by wind turbines, including the rare Griffon vultures and the even rarer Egyptian vultures. One Spanish wind farm (Navarra) has the gruesome record of killing 400 Griffon vultures each year. How long can this go on for before the species is extinct? These figures are probably an underestimate when compared with statistics published in December 2002 by the California

Energy Commission showing that bird deaths per turbine per year were as high as 309 in Germany and 895 in Sweden. Twice as many bats are killed when their lungs implode due to air pressure changes created by the turbines.

Because wind farms tend to be built on uplands where there are good thermals, they kill a disproportionate number of raptors such as the wedge-tailed eagle (Australia), golden and bald eagles (USA), Egyptian and Griffon vultures (Spain) and white-tailed eagles (Norway). Birds cannot biologically adapt to being sliced and diced by turbine blades that travel at over 200 kilometres per hour at the blade tip.

Biology can do some remarkable things but it is difficult to learn, evolve and quickly adapt flight patterns after being sliced and diced at 200 kilometres per hour. Loss of habitat is the single biggest cause of species extinction. Wind farms not only reduce habitat size but also create zones that attract animals and then kill them. Birds see turbines and wind towers as perching sites and are lured in (especially as grass variations beneath towers attract more prey). This is a lethal attraction.

Since 1846, there have been only eight sightings of the white-throated needle tail in the UK. Some 40 birding enthusiasts went to the Hebrides in the summer of 2013 to catch a glimpse of this brown, black and blue bird that breeds in Asia, winters in Australasia and catches summer in Siberia. The last sighting in the UK was 22 years earlier. The bird is capable of flying at over 160 kilometres per hour and was off course. The bird spends most of its time in the air, feeds on the wing on insects and in winter enters a coma-like state to preserve energy. The bird watcher crowd watched in horror as this rare bird was killed by the blades of a wind turbine. More wind farms are planned for the Hebrides. Again, the silence from the greens was deafening showing that the greens are not interested in the environment. Imagine the green cacophony if a coal mine had killed this rare white-throated needle tail.

Bat death mortality is huge, various studies worldwide return different numbers but the conclusion is the same: wind farms are catastrophic for bats. The Leibniz Institute for Zoo and Wildlife Research showed that bats killed by German turbines have come from places 1,500 kilometres or more away. In the name of green ideology, German turbines kill more than 200,000 bats a year. This is reducing bat populations across northern Europe. In the US, the death toll is as high as 70 bats per installed megawatt *per annum*. There are 40,000 megawatts of turbines currently installed in the US giving an annual kill of 2.8 million innocent bats. Not to worry. We can all feel morally superior because we are emitting less carbon dioxide into the atmosphere and the planet is being saved. From bats.

Bats reproduce very slowly, live a long time and are easy to wipe out. This is why bats are so heavily protected with many regulations and conventions, despite their transmission of deadly diseases (e.g. Hendra virus). Bats evolved to fly at night when there are fewer predators and have survived very well until the greens outsmarted them with their slice and dice and lung imploding wind farms.

Offshore wind farms hide their carnage in the sea. They kill sea and migratory birds and reduce habitat availability for marine birds (such as the common scoter and eider ducks). Wind turbines only last for about half their promoted design life but even in their short life, turbines are pushing some birds and bats towards extinction. The submarine acoustic resonance disorients whales and other marine life that use acoustics for underwater communication. The greens just don't want to know, maybe because they are so desperate to believe in "renewable" energy and destroy employment-producing heavy industry. I argue that it is not a state of denial. It is a wilful disregard for the environment and shows that greens are only interested in social engineering at the expense of our freedoms and the environment. "Renewable" wind energy poses a far greater threat to wildlife than climate change.

Why don't greens demonstrate about the bird and bat ecocide by wind turbine blades? If they were genuine environmentalists, they would. The wind industry is pretty good with covering up, burying corpses and conspiring with complicit ornithological organisations. And why not when there are zillions to be handed out in subsidies? The obsession with unproven human-induced climate change means that many greens are turning a blind eye to the ecological and environmental costs of "renewable" energy. They try to claim that human-induced climate change will create extinctions.

A little bit of knowledge would show the greens that the species they claim are threatened by human-induced climate change have already survived at least 60 glaciations and warm interglacials in the current ice age and sea level changes greater than their worst case projections. Climate change will not drive these alleged threatened species to extinction. Green ideology, ignorance and hypocrisy will. If you, Joe Citizen, wilfully killed birds and bats you would be prosecuted. Yet, the wind industry is immune from prosecution. If the mining industry killed birds and bats, there would be a public outcry, the offending company would be prosecuted and the operation might be closed down. Not so for the wind industry.

The years of environmental impacts studies, public hearings and Court cases that a mining company has to undergo to open a simple employment-producing unsubsidised operation are far stricter than those for inefficient wind farm operators. Because it is erroneously claimed that wind energy is good for the environment and saves carbon dioxide emissions, the wind power generating industry is able to damage the environment with no recourse and is paid self-regulated subsidies for doing so.

It would not be possible to smelt ores to produce metals to make your stainless steel teaspoon using wind power. Instead of stainless steel, if bone or shells were used for spoons, the environmental damage to fauna would be horrendous and we would have a good chance of being poisoned by a bacterium or virus hiding in the pores of bone or shell. If we want to use stainless steel cutlery to eat (or indeed anything made in the modern world), forget wind power. It's a practical disaster, an ideological contradiction and extremely damaging to the environment.

Solar power

Solar power has a great appeal to dreamers, Sun worshippers and the unbalanced. After all, we were all told by the greens that the energy from the Sun is pure, free and will go on forever. To take advantage of the warmth and natural light is, of course, the moral and enlightened way to use and conserve energy. This is true until one looks at the fundamentals. Solar power exists only because of subsidies and the mistaken belief that it reduces carbon dioxide emissions. It does not. Solar power, like wind power, actually adds to carbon dioxide emissions.

The US Department of Energy concluded that solar electric systems couldn't meet the energy demands of an urban community or industry. Large scale solar arrays, whether photovoltaic or solar thermo-electric are far too variable, unreliable, expensive and ecologically damaging. Solar can only make a minor contribution to the national power requirements. Solar power is not very efficient and the optimal figure used for incident radiation is 10 watts per square metre with an overall system efficiency of 5%.

Other factors affect the efficiency of solar radiation such as latitude, time of year, time of day and aerosols. There are also long-term weather fluctuations due to cycles of cloud coverage that can change the efficiency by 4%. Aerosols can reduce the efficiency by almost 30%. Furthermore, in remote areas, lack of regular cleaning off of dust and plant spores from the glass surface covering the photovoltaic cells can result in reductions of efficiency by up to 50%.

Solar cells (or photovoltaic systems) have been known since 1839. One would have thought that if they were to be a low-cost efficient competitive energy dense system for humanity, then more than 170 years of improvements should have been enough time to make solar power efficient. Apparently not. Notwithstanding, some greens argue that we should invest in solar power as the breakthroughs are just around the corner. What breakthroughs?

Solar cells do not create energy. They convert a little bit of sunshine into energy because electrons in silicon transfer from one state to another when hit by solar radiation. They must absorb (or release) the exact amount of energy to account for the energy differences between the different states. A silicon solar cell has a very narrow band gap (1.1 electron volts, eV). This corresponds to a wavelength of about 1,130 nanometres (nanometre = one billionth of a metre, nm) which is in the infrared part of the spectrum. If sunlight on silicon induces an electron to jump from one state to another, this transition releases a very small amount of electricity. If the broad spectrum of light strikes the solar cell, there is a quantum transition (1.1 eV, which is electricity) and the excess energy from the rest of the spectrum is wasted as heat. This is the limitation of solar cells.

This is why even the best silicon solar cells have an efficiency of not much higher than 10% and why we shouldn't wait around expecting huge efficiency improvements. In reality, efficiency is even lower because of light reflection and current leakage.

If the greens want to make silicon solar cells more efficient, then they need to pray to Gaia to make a few new laws of physics.

New developments

Surely new developments with metals, metalloids and super conductors are just around the corner and these will save the day. Cells of higher efficiency require the use of exotic, rare and poisonous elements such as germanium (Ge), gallium (Ga), indium (In) and cadmium (Cd). There are just a few practical problems. There are no germanium, gallium, indium and cadmium mines in the world. These elements are by-products from the zinc (Ge, In, Cd), aluminium (Ga) and tin (In) smelting and refining industries. To produce 1% of the US electricity requirements from a germanium or gallium solar cell would require three times the planet's annual production of germanium and twenty times the world's annual production of gallium.

The zinc (for germanium) and aluminium deposits (for gallium) are yet to be discovered and, if they were, it would not be economic to produce massive excess amounts of zinc and aluminium just to provide germanium and gallium. Zinc and aluminium are the two metals with the highest amounts of embedded energy and, to create small volumes of cheap solar power, astronomical amounts of conventional base load energy would be required.

If more gallium were to be produced for solar cells, the already marginal aluminium industry would have to greatly increase production and flood the market with massive quantities of unwanted aluminium. There is, after all, a limited market for cooking foil. Furthermore, these elements are far more expensive to produce than silicon (the second most abundant element on Earth). Just to support these germanium or gallium solar cells, 17% of the US annual cement production would be needed. And, to make cement, limestone needs to be burned and carbon dioxide is released to the atmosphere.

However, solar power is an elegant solution for small power needs (hundreds to thousands of watts) in remote areas where energy can be stored in conventional batteries. Examples are lighting, telecommunications systems, navigation beacons, recording equipment, marine buoys and satellites. The economics and inefficiency can be justified, especially as the maintenance requirements are modest. For example, telephone booths in outback Australia well out of range of mobile phones are powered by solar cells.

The Earth's atmosphere is about 1,000 kilometres thick and is divided into five layers. We live in the 12-kilometre thick troposphere. Above the troposphere are the stratosphere (12 to 50 kilometres altitude), the mesosphere (50 to 80 kilometres altitude), the thermosphere (80 to 700 kilometres altitude) and the exosphere (700 to 1,000 kilometres altitude). At the edge of the atmosphere, 1,367 watts per square metre reaches Earth from the Sun, while on average only 47% of this reaches the

Earth's surface. In ideal conditions when it is midday in the tropics, about 6% of incident energy is reflected and 16% is absorbed.

If a solar panel is to generate maximum electricity, conditions must be ideal (i.e. middle of the day, clear sky, low latitude). The maximum incident solar radiation value is 1,000 watts per square metres. It is claimed that an off-the-shelf solar panel will produce 110 watts for a one metre square panel. This is an efficiency of only 11%. The maximum incident solar radiation value is 1,000 watts per square metre. To produce this peak of 1,000 watts, 9.9 square metres of panels would be required. Solar panels can be flat or can use concentrating collectors and solar trackers to give maximum incident solar radiation for a longer time. In high winds, these concentrating collectors and trackers have to be closed down. Long-term measurements show that the average radiation is 125 to 375 watts per square metre could deliver 3 to 6 kilowatt hours per square metre per day. So far, so good. Now for reality.

Assuming a very generous optimistic efficiency for the average solar panel of 15%, an average of 19 to 56 watts per square metre would provide only 0.46 to 1.35 kilowatt hours per metre per day at an average of 0.61 kilowatt hours per square metre per day. The glass cover that protects the solar cells reduces the efficiency to 13% and further system losses of 7% can be expected due to localised conditions and the conversion of direct current (DC) to alternating current (AC) which we use in our domestic life. The US Department of energy calculated that solar panels have a 10.27% efficiency which equates to 4.25 kilowatt hours per square metre per day. Solar power advocates quote the maximum radiation value of 1,000 watts per square metre and don't really worry about those trivial little inefficiencies.

Logistics

There are a few other unimportant trivialities. A 1,000-megawatt nuclear or coal-fired power station occupies an area of 30 to 60 hectares (75 to 150 acres). A 1,000-megawatt solar power station would have to produce enough energy for an eight-hour day plus reduced energy production for the remaining 16 hours. The area required is 55.5 square kilometres. To do this with an efficiency of 10.27%, the area of solar panels, space between panel to prevent shading and maintenance roads would have to be 128 square kilometres (50 square miles; 12,800 hectares). All plants (and hence animals) would be removed from this 128 square kilometre area just to produce ideological inefficient unreliable electricity. This is certainly green environmentalism at its very best.

To build a 1,000-megawatt solar power station, it is not only the solar cells that are needed, there are structural supporting materials, concrete foundations, transmission systems, access roadways and a dispersed area of collectors and AC converters. The amount of material required is huge. Furthermore, massive earth works as site preparation would have to be undertaken.

To produce the 35,000 tonnes of aluminium for structural support only 12,777,950,000 kilowatt hours of electricity is needed which is now embedded energy and 735,000 tonnes of carbon dioxide is released during the process of smelting and refining to produce the 35,000 tonnes of aluminium. Some 75,000 tonnes of glass is required to cover solar panels and the manufacture of this glass releases only 45,000 tonnes of carbon dioxide into the atmosphere from degassing the components of glass. An additional 214,500 tonnes of carbon dioxide are released into the atmosphere to provide the electricity to melt the glass components. The embedded energy in the glass is 661,425,000 kilowatt hours of electricity.

Some 600,000 tonnes of steel are required for the 1,000-megawatt solar farm and electricity used for the manufacture of this steel would release 3,901,576 tonnes of carbon dioxide and the blast furnace would release 1,218,000 tonnes of carbon dioxide into the atmosphere. The embedded energy in the steel is 9,070,000,000 kilowatt hours of electricity. For wiring, some 7,500 tonnes of copper would be required. It has embedded energy of 529,507 kilowatt hours. To make this copper, 13,500 tonnes of carbon dioxide would be released from the smelter and the electricity used to run the smelter would release 494,200 tonnes of carbon dioxide into the atmosphere.

The two million tonnes of concrete for footings would release 2,706,885 tonnes of carbon dioxide for the electricity used and 360,000 tonnes of carbon dioxide from burning limestone for cement manufacture. The embedded energy in the concrete is 6,249,000,000 kilowatt hours of electricity. Just for these components alone, some 9,688,661 tonnes of carbon dioxide would have to be released to the atmosphere and the 1,000-megawatt generator would have to work at an efficiency of 10.27 for over 24 years just to pay back the 20,804,705 megawatt hours of embedded energy. Unless solar power generation is very heavily subsidised, it is clearly uneconomic. Why should the average punter pay for environmentally devastating uneconomic subsidised electricity?

These are minimum figures because of carbon dioxide emissions from road transport, site machinery and manufacture of other metals have not been calculated. Nor have the use of vehicles for maintenance. Over 100,000 truck loads of concrete would have been delivered by diesel-powered trucks emitting carbon dioxide and particulates. For example, there is a huge amount of energy needed to convert silica (quartz) to 99.99999% silicon for the photovoltaic cells. To make a very thin silicon wafer, 25 grams of carbon dioxide per square metre is released during manufacture of solar cells. Just to manufacture the silicon, 3,200 tonnes of carbon dioxide would have been released into the atmosphere. Far more carbon dioxide would be released during the packaging and transport of these cells.

There are numerous poisonous, flammable and hazardous chemicals used in the manufacture of a silicon solar panel. For example, arsenic, cadmium and lead are used in solders and a cleaning fluid for silicon manufacture is sulphur hexafluoride, a greenhouse gas that is 25,000 times more powerful than carbon dioxide. A recent calculation showed that to manufacture a 2-gram silicon chip, 72 grams of chemicals, 1.6

kilograms of oil or coal equivalent and 3.2 tonnes of water are used. The real environmental cost for the manufacture and decommissioning of solar cells is not known but it is not pretty. And all this for a short life solar cell. When solar power was all the rage, numerous Chinese companies were established to manufacture and sell solar cells to credulous Western green-contaminated countries. These companies are now disappearing at a very rapid rate, as are the Chinese poisoned by the pollutants used to make solar cells. Presumably for the greens, saving the planet is worth the human cost of killing Chinese workers.

The use of huge amounts of energy and the release of at least 10 million tonnes of carbon dioxide into the atmosphere just to construct a 1,000-megawatt solar powered generator in order to save the planet from increased carbon dioxide into the atmosphere does not look like a good idea, especially as at least 128 square kilometres of plant and animal habitats would be destroyed and the energy produced would be inefficient, unreliable and costly. Furthermore, because nature does not co-operate with dreaming ideologists, coal-fired electricity would have to be used as backup and the coal-burning generators would just keep burning coal and emitting carbon dioxide.

If Australia were to meet a 20% "renewable" energy target by using only solar power, then some calculations can be made. For 20% of Australian electricity production (53,000,000 megawatt hours) at peak production, a collector area of 55.4 square kilometres per 1,000-megawatt peak would be needed and, with space between rows for shadow avoidance, maintenance and cleaning, the area would have to be 128 square kilometres (50 square miles) per 1,000-megawatt peak. In areas at high latitudes it would be larger. For total power generation of 53,000 megawatts, the area would be 2,850,000 square kilometres of cell area and a land area for the solar power facility would be 6,863,468 square kilometres. Australia's land area is 7,685,855 square kilometres hence for 20% of Australia's total energy to be provided by solar power, 90% of the land area would be used.

If 100% of Australia's power were provided by solar power, 34,317,342 square kilometres of land area would be required. This is four and a half times the current land area of Australia. Australia would then not be able to feed itself, would have no land area for employment-generating industry and the population would have to live on boats or underground. When critics of the greens claim that green schemes are not practical, they don't say how they are really totally off the planet. It must be remembered that the solar power plant would provide 1,000 megawatts during the peak time of the day and could not provide a continuous 1,000 megawatts 24/7 all year.

Solar power has a low capacity factor. In Germany, it is about 10%. Hence 10,000 megawatts of solar power capacity is needed to generate the same amount of electricity as a 1,000-megawatt thermal coal or nuclear power station. Furthermore, when the 10,000-megawatt solar power generator is producing its maximum of 10,000 megawatts, the grid system cannot cope and hence huge yet-to-be-invented energy storage systems are needed or the solar power station needs to be shut down. Now that's efficiency.

Solar power generation at night

Sunny Spain was touted as the perfect place for solar power generation. Spain spent a fortune on constructing solar and wind power generators with generous subsidies. Spain became so clever at generating solar electricity that it even managed to do it at night. Generating solar electricity at night? No, it is not the new physics.

It is because subsidies were so incredibly high that solar power companies could make money by illuminating solar panels with floodlights at night. The floodlights were powered by diesel generators. This is madness.

In a 2009 testimony to the US House Select Committee on Energy Independence and Global Warming, it was shown that for every green job financed by Spanish taxpayers, 2.2 jobs were lost. Only one out of 10 green jobs were in maintenance and operation of already installed "alternative" energy plants and the rest of the jobs were only possible because of high subsidies. Each green job in Spain has cost the taxpayer $750,000 and green programs led to the destructions of 110,500 jobs. Each green megawatt installed in Spain destroyed 5.39 jobs elsewhere in the economy and, in the case of solar power, 8.99 jobs were destroyed per megawatt hour installed. I'm sure those pushed into unemployment by green activism feel that they have made a sacrifice for a higher cause.

On 16 January 2009, during a visit to an Ohio wind farm component manufacturing business, President Obama stated: "And think of what's happening in countries like Spain, Germany and Japan, they're making real investments in renewable energy. They're surging ahead of us, poised to take the lead in these new industries."

Spain has since gone broke, part of which was due to the extraordinarily high cost of electricity and subsidies. Thanks to the greens. And the green activists feel smug and take the moral high ground because of their policies.

Costs

One square metre of a solar panel costs $750. Installation doubles this cost. For a 1,000-megawatt plant to provide electricity in the depths of winter, 3,230,000 panels are required at a bargain basement price of $4.83 billion. This is only for peak production of 1,000 megawatts at the optimal time of day. If the solar power station were to compete with a conventional coal-fired thermal power station providing 1,000 megawatts constantly with a load factor of 70%, capital costs for a solar power station would be in the order of $100 billion. The latest coal-fired thermal power station built in Australia cost $1 billion at current costs and over its 20 year life would consume $2 billion worth of coal. This, of course, assumes that a solar power station would last 20 years at peak efficiency. An optimist would give it five years, at best 10 years because wafers of silicon quickly reconstitute and become even less efficient.

There are claims that solar panels are becoming cheaper and cheaper. This may be so but is does not change the facts. Solar power is still far too expensive, too unreliable and too environmentally damaging. Whatever the cost of solar panels, solar power cannot compete without massive subsidies. A recent study in Germany showed that solar power is four times as expensive as power from a prototype nuclear reactor being built in Finland, which is also a particularly expensive design.

When wind and solar electricity can so easily be shown to be uneconomic, why on Earth should there be a "renewable" energy target? When it can be so easily shown that carbon dioxide does not drive global warming, why should there be costs and restrictions on employment-producing industries that emit plant food?

From any perspective solar power is too expensive, too environmentally damaging and cannot provide large-scale energy to an electrical grid system. The greens hope against all knowledge that a large-scale low cost method of storing electricity for days, months or years is just around the corner. Such technology does not exist and is not even on the horizon.

Storage systems

Various storage systems are currently used. Batteries are normally used for solar cells but these have great limitations. For high temperature solar concentrators, costly high pressure-high temperature vessels of superheated steam or molten salts can store energy. For low temperature domestic air and water systems, hot water tanks are used. Water has a high thermal capacity and heated water is an effective low-cost way of storing energy. It is already widely used in industry.

Hydro pumped storage works on a 24-hour cycle. It depends upon site development, reliable water supplies and huge elevated reservoirs. Water is pumped up to storage dams that later release water for generation of peak load electricity. There is a net loss of energy and the process is costly however this energy storage process is used more for convenience than for cost. If green advocates of solar power want to use the only tried-and-proven (albeit expensive) method of energy storage, then there will have to be many more rivers dammed for hydro storage. But wait, aren't the greens against building dams despite the fact they generate jobs, store domestic and industrial water, store water for growing food and store water for "renewable" electricity production? Energy storage is costly and inefficient. Greens can't escape the fundamentals: the cheapest, most effective and most environmentally friendly way to create electricity is to generate it by the tried-and-proven conventional methods when we need it.

That's only a small part of the story. The largest cost for a solar power station is not for the components, maintenance and construction of the plant. The highest capital cost is in the distribution and support systems for solar power systems that cover such a large area. Even if a completely revolutionary system of energy storage is invented, solar power systems still could not produce electricity as efficiently, cheaply and reliably as the tried and proven conventional base load generating systems. In the unlikely event that solar cells were made more efficient (which is not possible because of the 1.1 eV generated by light at 1130 nm striking silicon), the distribution and support costs would not change. There is no point in waiting around until some bright spark invents a more efficient system. It could be another 170 years. And solar power would still be subsidised, environmentally devastating, inefficient and unreliable.

Biomass power

Maybe a way to get energy for the mining, smelting and fabrication of materials for your stainless steel teaspoon is from biomass. Greens champion biofuels as a weapon against alleged carbon dioxide-driven global warming. They claim less carbon dioxide would be emitted than fossil fuel alternatives. As plants soak up carbon dioxide while growing, the combustion of biofuels simply puts carbon dioxide back into the air resulting in zero emissions. That's the theory. Let's look at reality.

The energy density of wood is much less than that of coal or gas. Upon burning, wood produces about half the energy of an equivalent weight of coal and releases an army of complex chemicals into the atmosphere. Some of these chemicals are toxic. Wood harvesting is far more expensive than coal mining and wood-fired electricity is twice as expensive as coal-fired electricity. It might be romantically blissful to sit in front of a fire drinking *glüwein* and eating a wood oven cooked pizza, but that is the preserve of the wealthy in the West who can afford to pollute and be inefficient.

Nevertheless, there is a push to change from coal- to wood-fired electricity generation on the false assumption that the burning of wood reduces carbon dioxide emissions. This push comes from organisations that have been captured by the greens such as the US Environment Protection Agency. It was assumed that carbon dioxide released from wood burning is sequestered by vegetation for the growing of more biomass. However, the European Environment Agency has claimed that there are actually increases in carbon dioxide emissions. Most of the carbon dioxide in the atmosphere has a short residence time, it does not accumulate in the atmosphere and is sequestered into the oceans as dissolved carbon dioxide, bicarbonate and carbonate. Life uses these forms of carbon dioxide for shells, corals and phytoplankton respiration. The solubility of carbon dioxide is inversely related to temperature and cool ocean currents that upwell to reach warmer waters release massive amounts of carbon dioxide.

It gets even sillier. Burning of coal also releases carbon dioxide. This carbon dioxide originally came from the atmosphere and by burning coal is being returned to the atmosphere. Surely the zero emissions argument applies to coal also. How do plants decide to use carbon dioxide from biomass burning and not carbon dioxide from ocean degassing, burning coal, diesel, petrol and gas, fires and volcanic and animal exhalations? Do plants have selective photosynthesis whereby they use carbon dioxide molecules emitted by environmentally friendly good humans and not carbon dioxide molecules emitted by sinners?

The erroneous assumption by the Environmental Protection Authority, the EU and others is that burning of biomass is "carbon neutral". It is not. However, because it has been bureaucratically deemed that biomass burning is "carbon neutral", biomass-fired electricity plants receive carbon credits, tax exemptions and subsidies from governments who try to claim that they are environmentally friendly. Nations and utilities are not required to count their carbon dioxide emissions from biomass burning whereas coal and gas generators are required to make such calculations that are used to impose additional taxation.

Green madness

The UK is committed by law to a radical shift to green energy. By 2020, the proportion of electricity generated from "renewable" sources is supposed to triple to 30% and with about 10% of total electricity generation from biomass. The only way to produce so much electricity from biomass burning is to do what people have been doing for thousands of years. Chopping down forests. And why? Because the EU rules deem that burning wood is "carbon-neutral".

The extreme of green madness is the conversion of UK's Drax power station from coal to wood. Drax, near Selby, Yorkshire, was the largest coal-fired power station in Europe, generating up to 3,960 megawatts of electricity. This required 36,000 tonnes of coal per day. But, Europe wants to reduce carbon dioxide emissions from coal burning. In order to reduce emissions at Drax, to burn wood to generate nearly 4,000 megawatts of electricity, "only" 70,000 tonnes of wood per day would be needed. There are only a handful of large mines that produce 70,000 tonnes of ore a day. This wood is to come from the US (North Carolina) and will be shipped as pellets over 5,000 kilometres across the Atlantic Ocean from the purpose-built Chesapeake Port in Virginia.

To convert the Drax power station from coal to wood pellets, the UK taxpayers will have to pay £700 million and the new wood-generated electricity will triple the cost of electricity. The Drax Group plc will be subsidised over £1 billion *per annum* by the British taxpayer for this green miracle. Last year, Drax received £62.5 million in green energy subsidies and this figure is set to triple as the amount of biomass burning increases. The UK government has decreed that electricity customers will pay £105 per megawatt hour for Drax's biomass electricity which is £10 more than onshore wind energy and £15 more than electricity from a new nuclear power station to be built at Hinkley Point, Somerset. The current market electricity price is £50 per megawatt hour.

To harvest some 70,000 tonnes per day from the other side of the world is no mean feat. After clear felling North Carolina's forest of maples, sweet gums and oak in the swamp lands, the wood is converted to pellets in giant energy-hungry factories. This pelletising process uses large amounts of energy derived from coal- and nuclear-fired electricity generators in the US. For the 20-year life of the Drax power station, 511 million tonnes of wood will be harvested using diesel equipment from trees in the US to provide expensive subsidised "renewable" electricity in the UK. The tonnage of wood harvested and transported is more than most large mines produce in their lifetimes.

Trees just cannot grow fast enough to feed wood-fired electrical generators and "peak wood" will be reached very quickly because the trees harvested take 60 to 100 years to re-grow. This North Carolina wood will be shipped to the UK using oil-fired ships, unloaded using diesel equipment and transported using diesel trains. The equivalent of 46% of the energy generated by the wood-fired Drax generator will be used to transport the wood. Because wood has a lower energy density than coal, Synapse Energy Economics estimated that wood burning emits 50 to 85% more carbon dioxide than burning coal. The harvesting, transport and pelletising will produce huge amounts of carbon dioxide outside the EU jurisdiction so this does not come into the EU equation.

Where is the environmental impact statement for the Drax power station? A UK wood-fired power station has just passed on its environmental impact to the US for short-term gain. Harvesting forests results in increased soil carbon dioxide emissions. With such massive harvesting of wood, plants and animals would enjoy massive habitat destruction. US environmental groups claim that these forests comprise some of the most biologically important forests in North America and that there are risks to wildlife survival and biodiversity (especially birds). And what about the otters and pileated woodpeckers that inhabit these swamps? They don't really matter as the EU and UK greens are saving the planet. It is habitat destruction that drives species extinction, not climate change. This is green environmentalism at its best. Destroy the forests and their animals in another country for the sake of feeling good at home.

Green activists used to demonstrate against the Drax power station burning coal. It was Europe's single largest carbon dioxide emitter. The company now boasts of its "environmental leadership position" and states that they are the biggest "renewable" energy plant in the world. Demonstrations by green activists have ceased. The passive start to the environmental movement in the 1970s was against harvesting forests. Now, in order to keep the ideological home fires burning, green policies have led to the clear felling of 70,000 tonnes of trees per day in North Carolina for burning in the UK.

The UK government estimates that by 2020, 11% of the UK's electricity generation will be from wood. Will greens have massive protests against biomass harvesting and burning? No. Air pollution will be worsened, UK electricity will be dependent upon a non-EU country and far more carbon dioxide will be emitted to the atmosphere than if gas was burned. And who will pay for these inefficiencies and subsidies? The poor UK worker. Not that there are many UK workers left now as a large proportion of employment age people in the UK, especially in the north, live off welfare. And who pays for the welfare? The poor UK worker. Again.

In the UK, wood provided about 33% of energy in the times of Queen Elizabeth I. In the times of Queen Victoria it provided 0.1% of the energy. Why? The Industrial Revolution needed far more energy than could be provided by the fastest growing trees and so coal was used. If wood had been used to energise the Industrial Revolution, then a land area one and a half times the UK's land area would have been needed. Coal is plentiful and has a high energy density. English forests started to grow back and in 2000, the forest area was three times that of 1900. The same too for all European countries. It was the use of coal that saved the forests, not the greens or environmentalism.

A 40-megawatt wood-fired electricity generator in Cassville (Wisconsin, USA) burns 1,000 tonnes of wood each day provided by 30 different suppliers. Eventually the wood will be harder to harvest, transport distances will increase and costs will rise. And all this for a mere 40 megawatts. We've seen this before when European forests were clear-felled for glass and iron manufacture in the Middle Ages, as described by Agricola. The 100-megawatt Picway coal-fired generator (Ohio) looked at converting to wood but could not find a reliable long-term supply of wood. It will close in 2015 when more stringent Environmental Protection Agency emission regulations take effect. Jobs will be lost.

In Virginia (USA), taxpayers paid $165 million to convert the Altavista Power Station from coal to biomass. The power station is owned by a private company Dominion Virginia Power. Why taxpayer funds should go to a private corporation is beyond me, especially as Virginians will pay a higher cost for electricity. Politicians promoted the conversion as a method to "help meet Virginia's renewable energy goal". The claim was made that the Altavista station would be using biomass that would otherwise go to landfill. However, the Department of Energy has shown that 65% of biomass-generated electricity comes from wood and so the rest is from waste.

The EU Council of Ministers has refused to cap the use of biofuels. Originally they wanted their 10% "renewable" energy target for transport to come from biofuels. This was then reduced to 7% but there was no agreement about this cap so Europe is left with 10%. This will cost European taxpayers €13.8 billion per year for a reduction in emissions of nine million tonnes. In November 2010, even Al Gore claimed that his advocacy of corn ethanol, which uses around 40% all corn produced in the USA (or 15% of the world's corn), failed to feed the hungry and is wastefully used in engines. What Gore did not say is that for every 1°C average increase in temperature in the US Corn Belt, productivity increases by 10%. For every 1°C average decrease in temperature, the latitude for growing corn shifts nearly 150 kilometres southwards. A cooling event would greatly reduce the available land for growing food. It's happened before, it will happen again.

In the EU, crop biofuels (e.g. vegetable oils, sugar beet, canola, harvest waste, wood chips, ethanol) have replaced 5% of fuel used in transport. If biofuels were burned solely for transport, then carbon dioxide emissions would drop by 59 million tonnes by 2020. However, the International Institute for Sustainable Development showed that deforestation, fertilisers and fossil fuels used to produce the required biofuels emit about 54 million tonnes of carbon dioxide hence only five million tonnes of carbon dioxide emissions are saved. This is 0.1% of the total EU emissions.

Furthermore, the carbon dioxide saved is just emitted elsewhere with the net effect leading to an increase in global carbon dioxide emissions. If the standard unbelievable climate models used by climate "scientists" are run, EU biofuel use will postpone a modelled slight temperature rise of 0.00025°C by 2100 by 58 hours. Even climate "scientists" must be able to see that use of biofuels is pointless. An area as big as Belgium is used for growing biofuels and a similar-sized area is used for European imports. Biofuel farmland uses as much water as flows down the Seine and Elbe rivers combined. All this for a saving of five million tonnes of plant food. European farmers now use fast growing trees such as poplar, willow and *Eucalyptus* for biofuels. These plants emit the toxin isoprene.

A Lancaster University study suggested that the EU's 10% target will cause an extra 1,400 deaths at a cost of £1 billion annually from isoprene poisoning.

Moreover, there is an additional cost for biofuel electricity to UK taxpayers of £6 billion a year. Each tonne of carbon dioxide that is emitted to the atmosphere costs the British taxpayer £1,200. By contrast, the EU's cap-and-trade system costs about £4 per tonne and the British are paying 300 times as much for their carbon dioxide emissions than people on the continent. To make matters worse, economic estimates show that to cut emissions of carbon dioxide by one tonne, there is about £4 in environmental costs. These estimates are, of course, nonsensical. It has yet to demonstrated that carbon dioxide harms the environment or drives climate change.

Furthermore, one might have thought the greens would encourage the emissions of carbon dioxide because it is plant food. By emitting carbon dioxide, we are fertilising the planet and helping green plants grow. There is now good satellite evidence to show that the recent small increase in atmospheric carbon dioxide has resulted in a slight greening of the Earth. In the atmosphere, carbon dioxide is a trace gas, the dominant greenhouse gas is water vapour and, without these two greenhouse gases, there would be no life on Earth. Light, water and carbon dioxide create plant material, this process of photosynthesis appears to be unknown by greens.

There is a moral argument. Land is being used to grow fuel and not food in a world where one billion people are hungry. Food prices have been driven up by heavily subsidised biofuel farms taking the place of food farms. However, although environmental, economic and moral arguments can be aired against the biofuels industry, it is a huge business. Big green vested interests are living off subsidies and tax concessions. The costs of climate policies to stop the alleged global warming are now globally about £1 billion a day. Wind turbines cost about 10 times the estimated benefits and solar power costs about 100 times the benefits. The snouts are well and truly in the trough.

What happens if carbon dioxide does not drive global warming as I argued earlier? All this money has been wasted rather than used in lifting people from poverty, creating employment or preparing for a real disaster. What happens during the next inevitable glaciation when ice covers much of the Northern Hemisphere? No EU directive could change orbitally-driven glaciation, there have been dozens in the past and the next is due. The UK, EU, Russia, China and Canada would not have the available agricultural land to feed themselves and could only survive if they have a large amount of high-density low-cost efficient energy.

The biofuels green dream is an environmentally damaging immoral nightmare.

Frack off

The slick water technology of many decades ago has been transformed from a somewhat crude technology into something far bigger and safer. It is fracking. Advances in this technology have resulted in a decrease in carbon dioxide emissions in the US. The decrease in carbon dioxide emissions did not come from greens. It came from engineers who had spent years obtaining qualifications and decades of research before inventing a process that created jobs, released less carbon dioxide into the atmosphere and reduced the dependency on foreign petroleum by the US.

Fracking is now the latest green scare campaign. But what do we really know about it? Greens object to fracking not because it is unsafe, but because greens are losing control of energy policy.

How is fracking done?

Vertical wells up to 4,000 metres deep are used as a pilot hole to measure traces of oil and gas in sequences of shale, siltstone and sandstone. During drilling of both vertical and horizontal wells, the bit is cooled and the well is sealed with a water-clay mixture, drill chips and the drill mud are pumped to the surface up the centre of the drill stem, chips are geologically evaluated and the drill mud is recycled. The porosity and permeability of chips is measured as is the rock type. A more detailed analysis of the rock types is measured by lowering a logger down the hole that records gamma rays emitted by rocks. The logger actually measures natural radioactivity. Shales are more radioactive than limestone or sandstone, as are the bricks in your home. In fact, if you live in a house constructed of granite not only do you live with relatively high radioactivity but your cellar could contain an accumulation of the highly radioactive gas radon. I write this because greens try to frighten us with radioactivity scares but don't put matters in perspective.

These interlayered beds of sedimentary rocks are not conventional reservoirs. Conventional reservoirs comprise very porous and permeable rocks in which oil and gas are trapped beneath impermeable rocks. In many conventional oil and gas fields, there are normally unexploited tight sequences (i.e. porous but not permeable layers) that contain hydrocarbons and these areas are now being re-evaluated. After all, the best place to look for unconventional oil and gas is in an operating or defunct conventional oil and gas field that already has infrastructure such as pipelines.

A curved hole is drilled off the pilot well and is eventually drilled horizontally for up to 3,000 metres along an oil- and gas-bearing sequence of interlayered shales, siltstones and sandstones. The direction of the horizontal hole depends upon the direction of stress in the layer to be drilled. The technology for extremely accurate navigation of drill bits to within a few metres in horizontal drilling is astounding and continues to improve. Both the lateral and vertical wells are cased with steel pipe. Sets of small holes are then blasted through the horizontal steel casing at the deepest part of the casing.

Fluids (mainly water) are injected at high pressure through the casing holes into rocks at depth from the horizontal steel casing to open natural fractures and to create new fractures up to 200 metres from the horizontal hole. Closer-spaced horizontal holes are used when the sequences of oil- and gas-bearing sedimentary rocks are especially tight. Because the process takes place at kilometres depth where rocks are under great pressure, these fractures need to be kept open with sand or ceramic balls otherwise the high pressures will close the fractures and stop the induced oil and gas flow. Sometimes, acids, nitrogen and carbon dioxide are added to the water for very specific purposes.

The hole is then blocked with a plug above the first set of frack holes, a new set of holes is blasted through the casing higher up the hole and fracking fluids are again pumped into the well to create another set of fractures. This process proceeds up hole until up to 15 to 25 rings of holes for fracking have been produced. A smaller diameter hole is redrilled inside the casing to drill out the plugs. Once the drill bit is removed, the hole expels the fracking water for storage on site. Over the lifetime of a well, which could be 20 to 30 years, the volume of a large swimming pool of water is used. Most of the water is recycled. This is the volume of water used by a city in a few minutes. Furthermore, the burning of the produced shale gas produces water and this is also used.

Fracking fluid can flow out of the hole for 7 to 15 days, after that a mix of fracking fluid with some gas and oil is expelled and, after that, only oil or gas flows out of the hole. Oil and gas flow at high pressure through the blast holes in the horizontal $5^{1/2}$ inch casing. Oil and gas can flow to the surface or, more commonly, need to be pumped to the surface using a small electric pump within the cased well. A good productive well will produce anything from 300 to 1,000 barrels of oil per day. Some gas contains a small amount of oil (condensate) and oil and gas are separated at surface.

Fracking is used to extract oil and gas from tight rocks hence it is called unconventional. By contrast, oil, gas and water are extracted from very porous permeable rocks in conventional oil and gas fields by pressure or pumping and it is only clogged wells or wells at the end of their life that need fracking to squeeze out the last few drops of hydrocarbons.

Rather than use narrow roads to move large volumes of people, we build multilane highways. Rock fracturing is analogous. It allows larger volumes of hydrocarbons to be quickly moved.

There are millions of holes already drilled every year for water, engineering studies, mineral exploration, oil and gas exploration and research. Drilling of holes into and through aquifers takes place every day and the process is well understood. Fracked hydrocarbons come from deep sedimentary rocks below the water table. Many areas have multiple aquifers varying from an uppermost perched aquifer to various layers of porous and permeable rocks that contain groundwater in the pores. Groundwater and fresh water aquifers generally occur at depths of less than 500 metres and hydrocarbon productive sedimentary rock horizons occur at depths of 1,000 to 4,000 metres. Ground water often naturally contains small amounts of hydrocarbons as gas because plant and animal material trapped in sediments decomposes as sediments are compressed to form rocks. This gas can migrate from deeper in the sequence or be trapped in the porous permeable rock that water later enters.

Above the fracked zone, wells normally have three steel casing pipes within each other. They are separated by cement. The casing stops water flowing into the well or the hydrocarbons from deep down contaminating an aquifer. If one casing pipe is corroded or mechanically fails, there are still a number of barriers to stop mixing of aquifer water with hydrocarbons. Contamination of an aquifer would result in hydrocarbon loss so there are good economic reasons, as well as environmental ones, why contamination is not allowed to occur. For example, in Texas, the whole process is monitored and regulated by the Texas Railroad Commission that, as one of their briefs, has a duty to protect aquifers. If greens are concerned about what is a very efficient and environmentally friendly process, then they should study engineering, spend years on the job and then invent a better process. They don't. They just complain.

The US fracking industry is centred in Texas. The workers eat, sleep and play in the same communities in which they work. They drink the same water, they have the same environmental concerns about water pollution and water waste and they want their children to grow up in an uncontaminated environment. The number of direct and indirect jobs created in the US by the unconventional oil and gas revolution, as estimated by HIS CERA Consulting, at the end of 2012 was 2.1 million. This is the wake up call for Europe and the UK. Jobs and cheap energy (which creates more jobs) have been created by fracking. The US Energy Information Administration has estimated that from 2007 to 2012, jobs in the US oil and gas industry increased by 40% and jobs in the remainder of the private sector increased by 1%. All these jobs are sustainable while the US oil and gas industry thrives. These jobs were created on private and State lands. In 2012, the production of oil, natural gas, natural gas liquids and coal on Federally controlled lands decreased. Jobs were lost.

Dangers of fracking

There is no debate about fracking. For almost 70 years, fracking (human-induced rock fracturing) has been taking place in the US and other oil- and gas-producing countries and yet fracking has not been in the headlines. This is because fracking is a low risk event that has occurred millions of times for increased and more efficient production of oil and gas from vertical holes. It extracts oil and gas from exhausted fields, from damaged wells, from wells clogged with waxes and mineral precipitates and from tight rocks (i.e. high porosity but low permeability).

In more recent times, it has received media and community attention because it is larger, closer to more populated areas, uses horizontal holes and shows the uselessness of wind, solar, wave and tidal power. Twenty years ago, only insiders in the petroleum industry would have known about fracking and now it appears that everyone is an expert on fracking. The technology, efficiency and costs of fracking are being improved at breakneck speed and major changes are taking place in a time period less than the life span of an Italian parliament.

There are five blatant untruths that get recycled by greens opposed to fracking. These are: it pollutes aquifers, it releases more methane into the atmosphere than other processes of gas production, it uses large amounts of water, it uses hundreds of chemicals and causes damaging earthquakes.

Fracking has been taking place for seven decades in hundreds of thousands of wells in the US and so there is enough information to assess these five claims. After a comprehensive study, the Environmental Protection Agency in the US showed that not one aquifer has been polluted by hydrocarbons from fracking. Not one. Zilch. Zip. Zero. After 70 years, not one aquifer has been contaminated. Far more aquifers have been contaminated from over pumping resulting in an ingress of seawater or contaminated surface water.

The methane-charged tap water highlighted in the green propaganda film *Gaslands* was natural and unrelated to fracking. This was known by the film director before film release and the director still chose to air the demonstrably incorrect story. It made good theatre but was unrelated to fracking.

A disgruntled biologist claimed that shale gas fracking released more methane to the atmosphere than coal. Study after study showed this to be wrong. This objection to fracking was aired because methane, the main component of shale gas, is a far stronger greenhouse gas than water vapour and carbon dioxide.

As for water used, fracking in the US uses less water than is used on golf courses *per annum*. Fracking consumes a total of 0.3% of the total US water use. Water pumped into wells for fracking is blown out of wells by high-pressure gas and oil, stored in tanks and dams, transported, cleaned up and pumped back into deep aquifers or used for agriculture. Although Texas has droughts, it is not due to fracking. It is due to a lack of rainfall.

There are claims that fracking uses hundreds of chemicals. The main chemical is water (99.51%) and the remaining 13 chemicals (0.49%) are found in your kitchen, garage and bathroom. These are citric acid (lemon juice), hydrochloric acid (stomach acid), glutaradehyde (disinfectant), guar (ice cream), dimethylformamide (plastics), isopropanol (deodorant), borates (soap), ammonium persulphate (hair dye), potassium chloride (intravenous drips), sodium carbonate (detergent), ethyl glycol (de-icer), ammonium bisulphite (cosmetics) and petroleum distillate (cosmetics).

I admit that one of the chemicals is very dangerous. It is water. More people have died from drowning than from any of the other chemicals used in fracking. And the greens try to tell us that such chemicals are dangerous. This shows either their lack of basic knowledge, that they are very economical with the truth or that they have a great concern about the dangers of water.

As for earthquakes, pull the other one. Fracking produces earth tremors that can only be measured with very sensitive instruments. These are not earthquakes. They are the same magnitude as earth tremors produced from wind, tides, traffic, loading and unloading of dams with water, landslides, mining, erosion, groundwater and oil extraction and subsidence. Wind turbines generate microseismic activity which in green ideology is acceptable whereas microseismic activity generated from fracking is not. The Earth enjoys tens of thousands of earth tremors each day and they tell us that planet Earth is dynamic. Each day there are a number of earthquakes, tens of thousands of times larger than tremors and these result from the pulling apart and pushing together of the Earth's tectonic plates. Most earthquakes are submarine and occur along the mid-ocean ridges.

To imply that fracking produces damaging earth tremors shows a complete ignorance of basic Earth processes. Maybe it is not ignorance but deliberate deception? Of course, there is a basic question. Why is it that the media do not challenge such claims? Is it because they have no basic knowledge or because they want a sensational story? Probably both. However, all major networks have science and environment journalists who know at least the scientific basics regarding earthquakes and should be able to grill those greens making such obvious erroneous claims. They don't. Why not?

Fracking risks in the US

It is not all beer and skittles in the US with fracking. Texas is booming. However, west of the Mississippi almost half of the land belongs to the Federal government including 48% of California, 62% of Idaho and 81% of Nevada. If the Department of the Interior adds 757 new species by 2018 to the Endangered Species Act as a result of green pressure, then the sue-and-settle activities of greens will result in a great reduction in the lands available for fracking in the most productive oil and gas fields. It is no surprise that the Obama alliance with the greens comes into conflict with the shale gas-driven economic boom in the US and Obama's legacy may well be protecting species such as the sage grouse resulting in huge economic cost to humans. Stranger things have happened.

Let's make the erroneous assumption that fracking is dangerous and hence it should be banned. This is illogical. In the US, 30,000 people are killed each year in vehicle accidents yet cars are not banned.

Europe and fracking

US shale gas is having an effect on energy generation in Europe. Coal has been displacing gas for electricity generation in Europe since 2009. And a wise move too. There is again, another golden age of coal for Europe. This is thanks to cheap US imports, expensive green policies and energy insecurity. The shale gas revolution in the US has resulted in electricity generators changing to gas and on-selling contracted coal purchases to Europe thereby creating lower US emissions of carbon dioxide (if this is considered important; I don't).

US coal in Europe is cheaper and more secure than Russian gas, Europe has not developed a shale gas industry because of the green lobby's shale gas blockade, Europe has reduced its nuclear power generating capacity, Germany plans to abandon nuclear power generation by 2022 and solar and wind have been subsidised expensive failures.

What is the more secure form of energy for Europe: coal from the US or gas from Russia? Western Europe already imports about 60% of its natural gas, mainly from Russia, Norway and Algeria. Much of the Russian gas is piped through Ukraine and Belarus where disputes in the past have shut the pipelines. The rest of the gas is piped through Slovakia, Poland and Czech Republic. The new Nord Stream pipeline takes gas directly from Russia to Germany. As domestic supplies dwindle and fracking bans stop local gas production, Europe is at risk. The gas import proportion is expected to rise to 80% by 2030. Russia is bringing its 2,400 kilometres South Stream gas pipeline on line and the Shah Deniz 2 project in Azerbaijan will provide gas to Europe in 2018-2019. It is clear that Europe needs a fracking industry to create energy independence from Russia, lower energy costs, stimulate heavy industry and reduce unemployment.

The end result of Europe reverting to low-cost US coal is a spectacular own goal for the European green movements. Greenpeace, World Wildlife Foundation and Friends of the Earth insisted that emissions from the burning of coal produce carbon dioxide emissions that will destroy the planet. These groups, responsible to no one, need to accept their role in creating costly inefficiencies, energy poverty and unemployment resulting from their quixotic illogical journey to a carbon-free future.

Those dreadful Americans did not sign the Kyoto Protocol yet the US carbon dioxide emissions decrease was not driven by legislation, signing of protocols or feel good conventions. It was driven by the free market in a dash for gas. US electricity generators are changing to gas without the inefficient and subsidised carbon pricing incentives provided by the EU Trading Emissions Scheme. The US, derided by greens as being too slow to respond to the climate change challenge, has the last laugh.

Europe is the exception to the benefits of cheap shale gas. Gas production in Europe is forecast to drop nearly 20% by 2017. Oil is a global commodity and, for most areas, gas is a regional commodity. By Europe not participating in new cheap clean energy, its citizens suffer. Germany is fourth on the list of highest industrial electricity prices in the world. In much of Asia, electricity is 30% cheaper for companies, in the US or Russia, it is cheaper by more than 50%. It will all end in tears for German industry.

The EU is considering imports of gas from the US, despite having 470 trillion cubic feet of potentially recoverable shale gas reserves comprising about 80% of the resource available in the US. France and Germany have banned fracking for fear of potential water contamination despite having large shale basins. By contrast, successful recent oil exploration in offshore Spain has led to an exploration boom for gas. The crisis in the Ukraine has had EU energy politicians having conniptions and trying to find the fine line between reality and ideology.

President Obama suggested that the EU rather than taking US gas exports, should diversify their sources of energy in order to make the EU less vulnerable to Russian blackmail and that they should open up fracking to develop its own gas supplies.

European commissioners are now considering abandoning a binding target for "renewable" energy by 2030. By 2030, the game will be over and EU energy policy will have converted Europe into a historical theme park for Chinese tourists. Greens are furious at even a mention of changing course well into the future and there is some opposition by the EU climate and environment commissioners.

The UK has been putting its toe in the water with the Prime Minister announcing that he wants better infrastructure, shale gas is the way to achieve this and by using shale gas, the economy could generate at least £3.7 billion a year as well as 74,000 extra jobs. The trade-off is that families living near fracking sites would get cash payouts and affected local councils could receive more of the taxes collected by Downing Street.

Meanwhile, protests against fracking continue in the UK as greens still claim that shale gas exploration is unsafe. Maybe the greens, with all their generous funding from governments and the credulous, could drill their own best practice shale gas well and find out from their own experience. Rather than demonstrating against possibilities of highly improbable events, they could use their own real data.

The cruel irony is that the EU provides massive subsidies for uneconomic farming (and uneconomic energy) yet bans are in place for a cleaner method for extracting wealth from beneath the land surface as fracking does.

Fracking and farming

There is a greedy industry that threatens the purity of groundwater, releases greenhouse gases, leaches chemicals into our life-giving precious soils, destroys the pristine beauty of the natural landscape, and destroys or reduces the habitat of our unique flora and fauna. It leaves bacteria- and virus-contaminated material over a large land area. This kills people. Roads are clogged with slow-moving heavy vehicles, noxious smelly fumes are released into clean air and workers are exploited. Their employer provides low wages and the safety record is appalling. A few make good money, those nearby have their property prices reduced and government subsidies slosh around to keep the wheels of this industry spinning. This industry is farming. Not fracking.

Farming needs a huge amount of space and, as a result of advances in fertilisers, genetically modified crops, insecticides, herbicides and farming methods, the land area *per capita* of mouths to be fed has decreased. The forests and suburbia have expanded. In the UK, farms contribute to 0.6% of the GDP, employ 3% of the work force and use 80% of the land. If mining or some other heavy industry took up the same land area as farming, then there would be an outrage. The UK car industry produces about five times the wealth as the farming industry and only uses 0.0000003% of the land area. In a mining country such as Australia, the area of mining leases is less than the area of hotel car parks. In twisted green logic, it appears that food production is good yet the energy, metals and chemicals industries required to produce this food are evil.

Food production is even more honourable if it is "organic" (whatever that is), despite the fact that "organic" food production uses far more land per unit of food output than modern enlightened efficient farming. However, the best estimates for the productive capacity of "organic" farming are that it is 75% as productive as conventional farming and probably more like 40% after fuel use and additional labour costs are considered. Furthermore, "organic" farming is a beneficiary of pest control on conventional farms and the pests don't even get the chance to reach isolated "organic" farms. "Organic" farms benefit directly from modern synthetic pesticides. "Organic" farming in Germany recently managed to pollute a large area, crops were contaminated and people died because of *E.coli*-rich "organic" sprouts. A Saxon company was growing "organic" sprouts and didn't test their water for *E. coli* which confirms the generally held view that "organic" companies and certifiers routinely fail to test for the safety of "organic" products, let alone their authenticity. Not that all "organic" foods are bad. I am a great lover of "organic water", a malt- and hops-flavoured solution that contains 95% water and 5% of the organic compound C_2H_5OH. One of the reviewers of this book was appalled at this admission and suggested, in the interests of saving water, that I should change to a single malt liquid that contains 38% of this organic compound.

Without mining and energy for fertilisers, machinery and food transport to population centres, there would not be enough food to feed the planet. No amount of green energy, "organic" farming or economic policies could even feed greens, let alone the rest of the world. Is that what the greens really want? Mass starvation. This has been recognised by Maurice Strong, the Canadian father of the Kyoto Protocol. In his memoirs he let the cat out of the bag "glimmer of hope … the reduction of human population … to the point that those who survive may not number more than 1.61 billion people who inhabited the Earth at the beginning of the 20th century". Strong, one of the fathers of the global warming ideology, wants to kill off most of the world's population. What method of genocide does he prefer and who will be selected to depart this mortal coil? Presumably Strong and other green ideologues will remain as global rulers. This is the true face of green ideology. Are you one of those dreadful sinners that the greens want to do away with?

If shale gas were used instead, the beautiful English scenery would not be ruined by wind farms and people would not suffer from the effects of low frequency wind turbine noise. A true environmentalist with an eye on the bottom line would support the exploration and exploitation of shale gas. And this energy would be high density for 24/7, not just when the gods of wind such as Aeolus, Zephyr, Notus and Eurus decide to perform. With so many derelict and high cost industries and so much unemployment in northern England, there are moral, economic, environmental and political reasons why the UK should embrace shale gas immediately. Shale gas gives the unemployed hope of a better future, the greens don't. Shale gas is good for the environment, the greens' wind and solar farms are not.

Free and dodgy markets

The same green groups in the US that fought construction of the Keystone XL pipeline from Alaska to the rest of the US are now trying to prevent the US from exporting natural gas from the proposed Cove Point, Maryland liquefied natural gas export terminal. The greens' opposition to exporting natural gas is not based on fact but an ideological aversion to any traditional fossil fuel energy source. At the time when Russia's President Putin is using his country's natural gas wealth to support aggressive foreign policy, this groundless activism must be questioned. Who do these greens serve? A previous communist leader in the Soviet Union would have called these greens "useful idiots".

The EU's unilateral disarmament over energy is exacerbated by backing away from nuclear power and refusing to frack. They have left themselves with no other choice than Russian gas. The greens' biggest triumph in the EU was Germany's transition to "renewable" energy. There are plans to close nuclear power stations. Coal mines closed and coal-fired electricity generators began to be phased out with substitution by wind and solar. The greens, a noisy minority, now control German energy policy. This policy is a long-term bonanza for the Russian state gas company Gazprom because Germany cannot depend on wind and solar electricity and will have to buy more Russian gas to power industry and to keep the lights on. If I were a Gazprom executive, I would be funding noisy EU anti-coal, anti-nuclear and anti-fracking green groups in order to establish long-term markets for my Russian gas.

In September 2005, German Chancellor Gerhard Schröder and President Vladimir Putin of Russia signed an agreement on behalf of their countries to build the $4.7 billion Nord Stream pipeline to bring Russian gas directly to Germany. The German government guaranteed to cover €1 billion of the Nord Stream costs should Gazprom default. Ten days later Mr Schröder and his party lost the election leading to his resignation from politics. Within weeks of his departure from politics, Mr Schröder accepted Gazprom's nomination to head the shareholders' committee of Nord Stream AG (i.e. chairman of the company). Schröder's Social Democratic Party was in a coalition with Germany's Greens. Who knows what goes on in the murky world of Russian gas, EU green groups and fellow travellers?

The utopian dream of the EU powered by "renewables" has become a nightmare. Even ex-Chancellor Schröder, under whom the "green energy revolution" started, is now calling for policy revision, warns of damaging and unachievable target for "renewable" energy and carbon dioxide reduction and now advocates lengthening the life of German nuclear power stations. I doubt if this is because of common sense. Previously, the UK and Germany were united in the belief that climate change would be the new mobilising target to save centre left political parties. However, the voters have seen rising costs, unemployment, environmental degradation and regulation. They have become sceptical of human-induced climate change and are rightfully punishing the centre left political parties for creating such a mess.

The only country that has had a substantial effect on decreasing carbon dioxide emissions did not sign the Kyoto Protocol and used the free market to promote a form of fossil fuel energy that produces less carbon dioxide upon burning. This is the US and much of the re-growth of the US after the global financial crisis of 2007-2008 was in part due to fracking for the production of shale gas. Did anyone in the EU, the UK or Australia take notes? If the greens really think that a slight increase in plant food in the atmosphere will destroy the planet, then green activists should shout from the hilltops that fracking is the only tried and proven unsubsidised method of reducing global carbon dioxide emissions.

They don't. Why not? It is pretty simple. The greens' objection to fracking is not that they believe it cannot work or that it is environmentally damaging, it is that they fear it does work and that they will lose control of a nation's energy policy as unelected green activists. The US has shown the way. The US is drilling its way to energy independence from the oil-producing nations. By producing shale gas from fracking, there are hundreds of years of low-cost, low-carbon fossil fuels. That would be the end of the greens' "renewable" energy ideological dreaming.

Who pays the anti-fracking greens?

If I were an oil or gas executive from Saudi Arabia, Iran, United Arab Emirates, Kuwait, Qatar, Mexico, Venezuela, Nigeria or Russia, I would be pouring cash into the green movements to stop fracking and indigenous shale gas and tight oil production around the world (especially Europe). It makes good economic and political sense. Because wind, solar, tidal and all sorts of other "renewables" are totally incapable of satisfying the energy needs of an industrialised country, major petroleum producers provide the industrialised countries with their required energy and make such countries dependent upon petroleum. Who is to say that such funding of green movements does not occur? At many local occupancy and green protests, the agenda is driven by those who are not locals but who breeze in for their exciting protest day before going on to the next protest somewhere else in the country.

Do the greens really want to create jobs or are they more interested in destroying industry and putting the average punter out of work? The behaviour of the greens shows that they are not interested in the environment. Their hare-brained schemes show that they have no practical, financial, engineering or scientific knowledge. Some might say they are ignorant, others might claim that they are hypocrites. I say they are both. Greens are not an environmental group, they are bullying political thugs who are interested in power, control over every aspect of our lives, the taking away of simple freedoms and the resultant destruction of your job.

Those of us who are true environmentalists want more jobs and cheaper energy, but not at the expense of despoiling the atmosphere, soils and waterways. We are the majority. We are not noisy. Public policy debates are not won by one side being quiet. It is the noisy greens that want unelected totalitarian power over the silent majority without the bother of the ballot box. This is not helped by politicians who oil squeaky wheels rather than presenting the logical moral, economic and political case for low-cost, efficient, reliable, employment-producing energy.

We environmentalists are aware that 3.5 billion people in the world lack adequate access to energy, that every eight seconds a person dies as a result of energy poverty and that there has been a massive unnecessary increase in household electricity costs in the Western world over the past five years. In Australia, the increase is 110%. These figures have been provided in 2013 by the International Energy Agency World Outlook, the World Bank and the CIA World Factbook.

The case against the greens is a moral case. Another great moral issue of our time is should we be lied to, intimidated and ridiculed just because we have a different opinion from a noisy minority of unelected greens? Why should we have expensive energy that creates unemployment when there are tried-and-proven methods of creating environmentally friendly cheap energy and employment?

3

KING COAL

No coal, no dole

Coal saved the Western world from poverty. As a result of the use of coal in the UK, Europe and USA in the 19th century, the benefits of economic development were spread from a small wealthy elite to the broader community. Coal raised most low-income earners out of poverty. Western coal-producing nations are now so wealthy that taxes can support eco-activists on the dole to put people in the coal industry out of work.

A stainless steel teaspoon is the cheapest and safest way of getting some food into your mouth. This is a luxury as most people on Earth don't have enough food and most Westerners don't know how lucky they are. Eating, energy, a roof over your head, peace and security are all taken for granted. But it is not luck. It is the result of thousands of years of innovation, experimentation and knowledge capped off with the foundations of Western civilisation derived from the ancient Greeks, Romans, Christianity, the Reformation, the Enlightenment, the coal-driven Industrial Revolution and democracy. The greatest dangers facing us on planet Earth today are probably not the rise of China, Islam or carbon dioxide emissions but global cooling and our own loss of faith in our inherited Western civilisation.

Without chopping down a forest, I challenge greens to show me how a stainless steel teaspoon can be made without coal or other fossil fuels. A modern coal-, gas- or uranium-fired power station using the latest technology can produce the electricity required to make a stainless steel teaspoon at a reasonable cost with minimal environmental damage.

As I show elsewhere in this book, wind, solar and bio fuel electricity generation just can't compete.

The good old days

It is claimed that times were good before the burning of coal to produce carbon dioxide emissions. We sat around arm-in-arm happily singing, picking berries in the forest and living off locally produced "organic" tofu. We lived "sustainably" and had a very small carbon footprint. Since then, sinful modern humans have degenerated into selfish greedy capitalists wilfully emitting dangerous carbon dioxide into the atmosphere. This is rather like the Christian view of the world with original sin, the fall from grace, absolution and redemption.

Nothing could be further from the truth. In previous centuries, we clear felled forests, killed everything that might give us protein or was a perceived threat and polluted everything we touched. We died at an early age, normally from a big bacterial blast. Peace, longevity and security were not the natural state of affairs.

The greens want us to go back to this alleged "sustainable" life and have us using "renewable" energy. I don't. I am old enough and have lived in the "good old days" in a semi-rural environment; and it was not that long ago. In every way, the modern world is far better than the "good old days".

On the land, solar power grew our crops, domesticated animals were vegans and ate grass and crop stubble. Hens survived on kitchen scraps and provided eggs. Sometimes they wouldn't. A large vegetable garden was necessary for survival. If we had an excess of one crop, we would share and swap with others. There were many worse off than us. Sheep were used as lawnmowers. The vegetable garden and fruit trees gave seasonal foods, gardens were fertilised with waste, farm animals had to be fed whether they were working or not, rabbits were a good source of protein from the bush and a few bob could be made selling the skins.

For weeks in spring, we had a glut of cabbage, peas and beans. In summer, we had a glut of corn and passion fruit. We could not afford to be vegetarians or vegans, a luxury embraced by a wealthy few in the modern Western world. We were all vegetarians but once removed; we cooked and ate herbivores. Cooked meat is far easier for the body to metabolise than raw meat. Heating and cooking were initially with wood that had to be collected and chopped. It was only later that we had coal gas available at our homes for heating and cooking.

Reticulated electricity powered lights and a radio, there were no televisions, computers, iPhones or DVDs. Blackouts were common. There was no such thing as a portable radio. The family sat around the large upright wireless in the lounge room and listened to shows deemed suitable by our parents. This was the only room heated in the house. Some of us made crystal sets to listen to the radio. These required building a wire contraption, winding coils, using a galena crystal and cat's whisker and then slinging an aerial over the roof and trees. The world changed for the better when there was enough money to buy a germanium diode. When the weather was good, modern music banned by parents could be heard through an earpiece.

There was also no such thing as a toilet that could be flushed. There was an outhouse with a seat placed over a can that was collected weekly. During winter time it was cold and dark out there and in summer time, there were snakes, flies and an unforgettable pong. Of course, passion fruit and choko vines grew on the outhouse. Once a week, the dunny man came in his 40-piston truck to collect a full pan and replace with an empty pan. Sometimes the poor dunny man would slip in the mud and spill some of the pan contents on him and he was washed down with phenol.

If someone left a room, the light was turned off. I still do. Rather than use electric lights during the day, we were encouraged to be outside and we now pay the price with skin cancers. There were clothes for three occasions. Good clothes, winter clothes and summer clothes. Wardrobes were not necessary. Clothes beyond their useful life were used for cleaning cloths. Solar and wind power were used to dry clothes. Early in life my poor mother had to battle with a firewood-heated copper for washing and it was only later that we could afford a washing machine. No house had a clothes drier or kitchen dishwasher. It was a child's duty to do the wash up with a sibling hand drying crockery, cutlery, pots and pans. If clothes were torn, they were mended by hand or with an old Singer sewing machine. Many clothes were made at home from bolts of cloth, especially school uniforms.

Very little food was purchased and most was produced by the family. There were no takeaway food outlets, few restaurants and if one could afford a night out, a hotel dining room was the only place to eat out. Paper bags were used many times and no lunch paper bag was thrown out. Most consumer items had no packaging and spare string, cardboard, paper, ribbons, buttons, bottles, jars, tin cans, bolts, screws, nails and wood were all kept as there was always some future use. Old rope, chain and cable were highly valued. This was a necessity then, it is now called recycling and apparently recyclers have the high moral ground today. I have had enough recycling in my early years to get me a front row seat in Heaven. A phone call was a luxury, as was a ride in a car. An overseas phone call cost a fortune and had to be booked days in advance.

There was no air conditioning. If it was cold, more layers were put on or a blanket was used. If it was hot we stripped off. We adapted to diurnal and seasonal temperature changes far higher than the most catastrophic warmist temperature projections by just opening and closing windows. Food was preserved by salting or storing in an ice chest. Each week, the ice man would bring a huge block of ice for the ice chest although often in summer the ice block did not last a week. Of course, we could not order a block via the Internet. We had to eat the perishable food and wait until the next delivery. Later there was the kerosene fridge that emitted an unforgettable stink that filled the house.

There was no obesity as people walked everywhere and food was not abundant. There was no junk food. We ate what today would be called health food because there was nothing else available. We walked to a village shop, school, the train station, church, and friends' places. It was only later that we had un-geared poorly-braked primitive bicycles for faster and more distant travel. Holidays were rare and travel to a holiday location was by a steam train burning coal and then by bus. The unforgettable pleasant smell of a steam train never leaves your brain. Travel abroad was out of the question.

Daytime childhood entertainment was playing in the bush with knives and weapons, building forts and tree houses that fell down after heavy rain or a puff of wind, building canoes from a sheet of rusty galvanised iron with nail holes that let the water in far too quickly or painfully losing bark with billy cart races. Collecting treasures by scavenging from old building sites and hunting and fishing were exciting options during daylight hours. At night, as there was no television, books were read before dropping into the sleep of exhaustion. On wet weekends, a great pleasure of mine was to get the train into the city and spend the day at the minerals gallery of the museum. This was the only childhood I had, hence it was the best childhood I ever had. I didn't live in a shoebox, was not forced to walk on broken glass and only have fond memories.

This might sound romantic to some. It wasn't. We were rescued from this frugality by large amounts of cheap energy generated from coal. Some people had reticulated town gas made from heating coking coal that produced gas and left a coke residue. Gasworks could be smelt a kilometre away. The coke was also useful and could be used for home heating. In the 1950s in Australia, there was an energy revolution and there were fewer coal miners' strikes. Electricity generated from coal became cheap, plentiful and reliable. Diesel and petrol became cheaper and we could afford to buy an old second hand car. Electricity could be used for heating, cooking and refrigeration, lawns could be mowed by a machine that did not leave dung everywhere, there was less need for muscle power, there was more free time and foods that had to travel could be sampled. A major event was connection of the house to the town sewerage system.

It took a long time to evolve from the standard evening meal of mutton or fish (generally flake or shark) and three vegetables. Leftovers went into a stew. At times, tinned sardines or kippers were an alternative source of protein. Rhubarb, custard, junket, rice and tinned fruits were the favourite desserts when fruit was not in season. Breakfast was always porridge, bread and ice cream were made at home, margarine did not exist and bread was sometimes covered with dripping if there was a shortage of butter or if it was too expensive.

The last thing at night was to place a two-pint billycan at the back door and in the morning the milkman would fill it with unpasteurised milk. The first child to bring in the milk had the luxury of scraping off the cream from the top of the billy. Later, milk was delivered in pint bottles with an aluminium foil lid. It was an early morning race to beat the magpies who had taught themselves to pull off the bottle cap and suck up cream. In the late 1960s, bland pasteurised milk in cartons was available from supermarkets. It was not until I was in my 20s that I ate chicken, beef, shellfish and tropical fruits and it was not until my 30s that I could afford to travel abroad. Despite all of this, we had stainless steel cutlery.

These were frugal times. However, we had a good formal education and understood where everything we used and ate came from. Both frugality and good education have taken a back seat in most of the Western world. We were well off compared to war-torn Europe. The UK lived off New Zealand, Australian and Canadian food exports and, although the World War II finished in 1945, the UK had food rationing until 1954. There was widespread malnutrition in post-war continental Europe. Australia was the land of milk and honey for displaced European war refugees. I still have childhood memories of European World War II refugee workers snaring pigeons on public buildings as a source of their protein and wandering the streets looking for odd jobs.

Cheap energy certainly allowed me to evolve from a "sustainable" life to a comfortable life. We now have far more time, better food, more food, more diverse food, fresh water on tap (filtered if preferred), frequent cheap travel, self-regulating home heating and cooling, swimming pools, robotic vacuum cleaners and mops, automated sprinkler systems, subsidised health care, safe working conditions, satellite navigation systems, communication, digital books, portable communication devices such as phones and iPads, home theatres with surround sound and hard drives full of terabytes of music, movies and documentaries.

Society and economics allow us to spend our free time doing things we enjoy as opposed to being factory fodder or hunting and gathering just to survive. We now have assets and are far wealthier. We can now afford to go to sporting events, concerts and restaurants. We can now afford to be greens.

Don't give me the "good old days"; they weren't. No green is going to tell me that I should go back and live that frugal "sustainable" life again. I've done my bit in my childhood and it did not save the planet. Those greens that want this allegedly environmentally friendly romantic life are quite welcome to become retrogressive and do it themselves. Somehow, I doubt if they could actually live a "sustainable" life, especially as most greens are based in cities. With no knowledge or experience of such a life, I doubt whether they would be capable of living the "sustainable" life that I had. And if they want "sustainability" and "renewables", they should have to pay for it themselves and leave me to enjoy my hard-earned comforts of life.

Until I see the major movers and shakers in the green movement living as I did when I was a child, I will continue to regard them as hypocrites with no credibility. I wait for the time when I can go to consult the green oracle, living in a cave in the bush, living off the land and using no trappings of the modern world. Until then, get out of my life.

Better times

Since the Industrial Revolution, life expectancy for people living in the

Western world has doubled. Ready access to cheap energy allows for better living conditions and easier lifestyles. Coal was the fuel of the Industrial Revolution and steam was its power. There was also a transport revolution to haul coal initially along canals and then later along railway systems from pits to steel and cotton mills. These transport systems allowed workers to travel short distances, enjoy a holiday at the seaside and to broaden horizons. Towns were lit by coal gas, train timetables meant that workers needed watches and inventors flourished to capitalise on the technological advances that came with the Industrial Revolution.

Coal allowed a life of desperation to be replaced by a life of aspiration. In northern England, there were community gatherings, competitions, brass bands and outdoor activities away from the darkness of an underground coal pit. Workers sought a better education. Women and children working in underground coal mines were replaced by pit ponies that later were replaced by machines.

Coal has brought hundreds of millions of people over generations from poverty to prosperity, despite population- and economy-destroying wars. The green white Westerners are totally immoral telling billions of people in Asia, Africa and South America that they can't escape from poverty and develop to our standard of living using coal, gas and carbon-based energy as we have. This is green racism.

In today's world, people are better fed, better sheltered and better protected than in any time in history. Overall prosperity has increased in the last century, as has mine. The average person's standard of living has improved ten-fold over this period of time, as has mine. Cheap inexpensive transportation, a consequence of the Industrial Revolution, has revolutionised the spread of trade, services and ideas. Without international trade, we could starve. For example, massive local rains in France in 1694 resulted in the third consecutive crop failure. Some 15% of people in France starved although plenty of food existed elsewhere in Europe. At that time, France was "sustainable", lived off the land and had very little trade with other countries. A convoy of 120 ships left

Norway with grain for France. The convoy was captured by the Dutch and was recaptured by the French and escorted to Dunkirk. It was too little too late as there was still not enough grain to feed everyone in France. Starvation continued.

The World Bank reported that in 1981, 42% of people in the developing world had to live on less than a dollar a day. Some thirty years later, this percentage has been reduced to 14%. This is a huge change in a short period of time. Such a change is unparalleled in history. The greens' actions attempt to reverse this trend. That is a moral conflict for the greens. Their solution: ignore it.

The World Energy Outlook of the International Energy Agency shows that there is a near perfect correlation between global electricity from coal and gross domestic product. Consumption of coal grew by 5.4% in 2011. Asia leads the consumption race. The world is becoming a better place. The *per capita* food intake, longevity and wealth have increased whereas child mortality, disease and land area for food production have decreased. There is still much to do but for a long time the trend has showed that the planet is becoming a better place. The same authority also shows that 3.6 billion people still have no or only partial access to electricity and the quickest and cheapest way to produce electricity is from coal. If we were really serious about the environment and the welfare of the poorest people in the world, we would flood India, China, Africa and South America with coal-fired power stations.

The Copenhagen Consensus Centre and the World Bank show that the proportion of extremely poor people has more than halved over the last 30 years, from 42% of the global population in 1981 to 17% in 2010. In 1820, more than 80% of people were miserably poor. There are still 20% of the world's population who are illiterate. Although this sounds dreadful, some 70% of the world's population was illiterate in 1900. China has recorded the biggest improvement.

In 1950, South Korea and Pakistan had the same level of income and education. Today, the average South Korean has had 12 years of education, in Pakistan it has not yet reached 6 years. The South Korean *per capita* income has grown 23-fold whereas in Pakistan it was 3-fold. Illiteracy costs. In 1900, global illiteracy cost 12% of global GDP, it is now 7% and the estimates are that in 2050 it will be 3.8%. These are world problems, we cannot ignore them. However, the world is a far better place than it was 100 years ago and global problems are being solved. But not by the greens.

There seems to be a view amongst the greens that we face crises for which there is no solution. If we look at the long history of experimentation, initiative, intuition, creativity, inventiveness, science and engineering, we can see that the end product, for example, is a brilliant 18:8 stainless steel teaspoon. We should not fail to look at what has been done in the past to improve our lives.

We humans are very good at overcoming hurdles to our existence and this we have shown since we began to live in villages more than 10,000 years ago. When we humans are dealt a wild card from the pack, we are pretty good at adapting and improving. We humans live on ice sheets, in mountains, in deserts, in the tropics, at high latitudes and at the seaside. If the need arose, no doubt civilisation would find a way to live under water.

We have already adapted to live in a great diversity of climates and if there were a change in climate, it would not create any problems for the human race as we already live in areas that vary from -40°C to +50°C. But what is an ideal climate for humans? The climate that the Eskimo prefers might not be the same climate that a hunter in Borneo prefers. There is no ideal climate for humans and with energy, technology and innovation we have already shown that we can survive extremes. The world is a better place, increased affluence makes it an even better place and affluence enables us to solve any potential environmental problems.

One only has to travel to some African countries to see that environmental degradation results from poverty, not the inverse. I have been visiting African, Middle Eastern and Asian countries off and on for nearly 40 years. As these countries crept out of crippling poverty (despite political despotism), the environmental degradation decreased. It is only wealthy countries that have enough funding to address environmental problems, perceived or real. The best way to make the world a better place is to have more wealthy people and to have fewer greens investing other people's money on tried-and-proven failures.

The world is now feeding more mouths with less land (hence less forest has been cleared), consuming more calories and spending less income proportionally on food than previously. Since 1970, the rate of increase in the global population has slowed. As the GDP *per capita* in a country increases, the fertility rate of women decreases and the demands on the environment also decrease.

The China and India factors

We are living at a time in history when there are two nations with more than a billion consumers. They have had thousands of years of gripping poverty and now there is a rapidly emerging large middle class. The world is in the middle of the biggest industrial revolution ever seen. Hundreds of millions of people are moving from rural China to the cities. This is the greatest diaspora the world has ever seen. These immigrants to the cities want our Western standard of living. They have had a smell of it and there is no turning back. These aspirants use commodities and the statistics from China and India are mind-boggling.

It is immoral for greens to promote costly, inefficient, unreliable "renewable" energy to the rest of the world. This stops people in Africa, Asia, South America and India escaping from the grips of poverty and reaching our standard of living. As an absolute minimum, these people need clean potable water and cheap reliable electricity.

China

China has 19% of the world's population. It consumes 53% of the world's cement. Concrete is made from gravel, sand, cement and water and any modern industrial growth is underpinned by concrete because it is a strong, cheap and durable building material. The high *per capita* consumption of cement shows that China is building, modernising and growing. Cement is made from heating limestone and shale to a high temperature with the heat derived from coal. Because limestone contains 44% carbon dioxide and China's energy is principally from the burning of coal, it is no wonder that China emits huge quantities of carbon dioxide and by 2020 (or earlier), China will be the world's biggest carbon dioxide emitter. This is good news as it means more and more people in China are escaping poverty. We should also thank China for putting so much plant food into the atmosphere.

If China reduces its carbon dioxide emissions, then its growth would slow. Notwithstanding, slight changes in the growth rate of China will have no great effect on the world. China will not curtail its growth in response to moralising by unelected wealthy Western greens, they want the same standard of living as enjoyed by city-based Western greens. If Western greens want China to reduce carbon dioxide emissions as a moral imperative, then they are knowingly and hypocritically keeping hundreds of millions in poverty as a result of their ideology. China wisely does not listen to Western greens. Why should we sensible folk?

Over 400 billion-kilowatt hours per month of electricity is consumed by China. The United States Energy Information Administration projects that China will bring on over 450 gigawatts of new coal-fired capacity by 2040. The demand for coal in China is expected to double from 2011 to 2016. China is reducing domestic steaming coal production and is continuing to close small and inefficient coal mines. Chinese domestic steaming coal adds sulphur gases and particles to the atmosphere thereby providing serious pollution in industrial centres.

By substituting less polluting and higher energy Australian steaming coals for the high-sulphur high-ash Chinese coals, energy production can increase without a proportional increase in air pollution. Indonesia and

Australia are the leading suppliers of coking and steaming coal and it is expected that in 2016 both countries will each export 450 Mt of coal to China. China is also looking to impose local production limits for the special and scarce resource of coking coal. With reduced access to domestic coking coal, China's demand for imported coking coal will increase.

China consumes 48% of the world's iron ore production. About two million tonnes of steel is made in China each day from local and imported iron ore and coking coal. China also imports about 60 million tonnes of steel each year and imports the same tonnage of iron ore each month. China produces 11 times as much steel as the USA yet China is only four times as populous. The US grew during the Industrial Revolution in the 19th, 20th and early 21st centuries. China is now catching up.

This steel is used for buildings, construction and for the 18 million cars that China makes each year. It is also used in the expanding motorway and railway system. Over the next 12 years, China will build 40,000 kilometres of railways. It already has the world's fastest train and the largest high-speed network in the world. China's third west-east gas pipeline made out of steel will be 7,400 kilometres long and be completed before 2015. In order to keep the lights on and to make steel and other products, China imports 47% of the world's coal production. This coal is coking coal for smelting and thermal coal for electricity generation. Each week China builds a new large thermal coal fired power station. And the story keeps going for almost every other major commodity.

China also is the number one producer of wind and solar power generators. They are not silly; they don't use such technology themselves to drive their own industrial revolution. They sell them to fools in the West whose governments subsidise unreliable electricity generation and live off debt. Not only does the West become more inefficient and internationally uncompetitive with unreliable subsidised electricity, but the Chinese are laughing all the way to the bank. And the Chinese bank is getting bigger. China purchased 2,600 tonnes of gold in 2013 at a cost of $US104 billion. This is more than the world's annual production of gold.

The Chinese also have the world's largest stockpile of foreign currency reserves. China accumulated $US3.4 trillion in foreign currency reserves in 2011, the largest currency stockpile on planet Earth. It is ironical that the country that invented paper money made out of mulberry bark is shifting its paper money and bonds into bullion.

The Chinese also eat. There are more pigs in China than in the next 43 pork-producing nations combined. The list goes on and on and on. However, there is not only an industrial revolution in China, there is also a knowledge revolution. They have shifted from 14th to 2nd place in the number of published scientific research articles and have the world's fastest supercomputer. It makes not one *iota* of difference to the global emissions of carbon dioxide if Western countries reduce emissions. China will make up for the shortfall before you can say coking coal, thermal coal and methane gas. Even though there has been an increase in longevity in China, 50,000 cigarettes are consumed each second.

In 2011, the increase in China's carbon dioxide emissions was 200 million times more than in the UK. Whatever carbon dioxide emissions savings the UK, Europe, Patagonia, Iceland, Timbuktu or Australia manage to achieve, it will have not one *iota* of difference to global carbon dioxide emissions by humans. China, India and East Asia want a better standard of living and to do this they will emit increasing amounts of carbon dioxide.

In China there has been a 45% increase in national income, protein sources are more varied and agricultural productivity has increased. Wine is a measure of consumer affluence. China now owns wineries in France and the New World and, as well as making its own wines, imports wines from many parts of the world. This would have been unheard of 50 years ago. The area of China's corn harvested over the last half century has doubled, each harvested hectare has become more than 4.5 times more productive than 50 years ago. The 120 million hectares of land spared from land clearing is twice the size of France. No wonder Chinese forests have expanded more than 30% over the last 50 years. As affluence in China increases, the rate of population increase is falling.

The moral issue is not that of burning fossil fuels and putting carbon dioxide into the atmosphere; the moral issue is that the greens want to keep most of the global population in abject poverty. Who are we in the West to deny others in the world the same opportunities, longevity and standard of living that we enjoy?

I can just imagine green activists trying to tell the Chinese administration that they must stop bringing people out of poverty and lower their carbon dioxide emissions. Maybe the Chinese would not be as kind as the Russians were to Greenpeace activists who attempted to illegally invade a Russian drilling platform in the Arctic. Maybe the families of Greenpeace activists would have had to pay for the cost of a bullet.

Can the China boom continue?

There is no guarantee that the China boom will continue. If Western consumers have money, they can continue to buy Chinese goods and maintain Chinese growth. Debt is the order of the day in the West and the good times may not roll on forever. A period of global cooling could greatly stress China and the rest of the world. Food security is a high priority in China and, during a cooling event, frosts and snow would kill off crops at the emergent stage. It happened in 1816. Global cooling would greatly reduce agricultural production in northern China. They temporarily rely on oil imports, mainly from Iran. Half the oil currently used in China is from imports and imported oil is now being replaced by oil created in local coal-to-liquids plants. China has reserved monstrous amounts of coals for future coal-to-liquids production. China is using coal from other countries while preserving their indigenous reserves.

China has probably about one trillion tonnes of coal resources and 40 billion barrels of crude oil. The EU bankrolled the building of wind turbines in China and 30% are not connected to the grid. The EU also partially financed the building of dams for hydro electricity and agriculture with "renewable" energy funds. Green activists would have stopped these dams being built in Europe, UK, North America and Australia. China has fast-tracked its nuclear power generating industry. China is almost energy-independent and has made big investments around the world to acquire mineral commodities while preserving local reserves. There are large reserves of some strategic commodities such as tin, tungsten and rare earth elements in China. No other country has significant reserves of these commodities. The US has a strategic stockpile of metals that are not mined in North America. Most other countries don't.

While the green activists in the West are forcing wind and solar farms to be built, China has not bothered to connect 30% of their gifted wind farms. China is building dams for hydro electricity and food production while the West is not. China is boosting its coal-fired electricity capacity and using gas for other purposes whereas the West is doing the opposite. Most Western countries have no coal-to-liquids programs and modest nuclear programs whereas China is fast tracking these energy systems. While green activist pressure in the West has led to destroying, ignoring or not using efficient energy systems, China has been racing towards long-term self-sufficient efficient energy whereas the West is looking increasingly exposed.

China will soon finish its strategic stockpile facilities for crude oil and will have the industrial, financial and military might for war against Japan and possibly Vietnam, Philippines or the US Pacific bases. In the distance, there is already the faint sound of sabres rattling. This may be a diversion from problems at home resulting from a slowing of growth, global cooling, problems of food scarcity and an opportunity to redress what is seen in Chinese eyes as a century of humiliation.

Indian growth

India's population has doubled since 1960, its national income has increased 15 times. Indians eat more and better than their counterparts in 1960. Since 1960, Indian forests have increased by 15 million hectares, less land is used for agriculture and she now exports food in contrast to former times when it imported food. India is a success story of chemical, biological and mechanical innovations arising from the Industrial Revolution. The move of Indians and Chinese to cities has actually resulted in a better rural environment because fewer people are now dependent upon the forests for food.

India will build an additional 75 gigawatts of electricity in the next five years. This will require the burning of 2,509 million tonnes of steaming coal, some of which will come from Indian coalfields and most of which will be imported. Thermal coal demand in India is expected to outpace production by at least 150 million tonnes within five years. The increased number of blackouts in India shows the need for more coal-fired power stations, increased coal imports and an improved power grid. Multiple new port projects are underway to enable increased imports of steaming coal. In terms of million tonnes of oil equivalent, in India coal dominates over oil, natural gas, nuclear, hydro and "renewable" energy. Globally, coal is the world's fastest growing fuel with coal growth from 2001 to 2011 of 56%, hydro 35%, natural gas 31%, oil 13% and nuclear 0%.

The same growth we see now with China and India was seen in the US during the Industrial Revolution. Between 1860 and 2010, the population grew nine times and the gross domestic product 130 times. Corn production rose 17-fold yet more land was planted with corn in 1925 than in 2010. This is due to better farming methods, mechanisation, fertilisers, insecticides, herbicides and genetically-modified crops. The volume of timber standing in US forests has increased appreciably from 1952 to 2010. None of this has been done by greens.

Where is the coal?

At present, the US holds 27.6% of the planet's coal reserves, Russia has 18.2% and China 13.3%. Coal resources are unknown but are orders of magnitude higher. Because coal is present in terrestrial sedimentary rocks that drape over most countries, almost all countries have resources and reserves of coal. Australia, India, Germany, Ukraine, Kazakhstan, South Africa, Serbia, Colombia, Canada, Poland and Indonesia also have substantial reserves and more coalfields are being found every year. The state-run Coal India Ltd will try to raise production by another 180 million tonnes giving a total production of 615 million tonnes in 2016-2017 compared to 435.5 million tonnes in 2011-2012.

Many nations keep coal in their core energy mix as a national security consideration and restrict export of it. Coal also provides more bang for the buck compared with wind, solar, biofuel wave or tidal energy. Despite green propaganda and wishful thinking, coal is not about to die. Coal will continue to thrive because it does what no other energy source can do. Coal is cheap and has every possibility of getting cheaper although top quality coking coal is becoming harder to source.

The UK was a significant coal producer. A century ago it produced 292 million tonnes of coal *per annum*, now it is less than 10 million tonnes. There is no shortage of coal reserves in the UK, production costs have risen greatly and the EU has put its green bureaucratic nose into the internal affairs of the UK energy industry. Furthermore, in 1999 the UK was at peak oil production of 2.9 million barrels per day. In 2012, it was 1.9 million barrels per day. The UK now imports almost all the fossil fuel it burns and is greatly exposed to the vicissitudes of the energy market. The lights could very easily go off in the UK. And stay off.

Australia is a large coal producer but its costs are rising. Coal is the second biggest export (after iron ore) yet more than half of Australia's coal mines have production costs above the global average. With capital costs in Australia two-thirds above the global average, Indonesia has now overtaken Australia with steaming coal exports to China. Five years ago, Mozambique and Mongolia were not involved in coking coal exports. Both countries are now starting to export it and Botswana is waiting in the starting blocks to be a coal exporter. US coal exports are set to rise dramatically in the next decade.

In response to noisy political pressure, we might decide to reduce emissions of carbon dioxide. The only effect will be to make our industries uncompetitive and increase our cost of living. This has already started to happen. Environmental carping about coal and carbon dioxide is unrelated to reality. If one country for some bizarre reason decides to stop exporting coal to save the world from a speculated carbon dioxide-driven global climate change, another country will take up the slack and the markets. Once markets are lost because of unreliable supply, trust is lost and markets cannot be won back. And nothing would happen to global climate by one country reducing its carbon dioxide emissions. However, export earnings and jobs would be lost. Forever.

There is an international market for coal and most other materials. Some greens have an objection to international trade and finance, not for environmental reasons but for political reasons. For thousands of years there has been international trade. There are numerous examples. There was Neolithic trade of obsidian from the Greek island of Melos all over the Mediterranean and inland into what is now Turkey. Persian bronzes contain copper from Esfahan (Iran) and tin from southern China.

The Romans had a large integrated market and a finance system for this market, as shown by the Sulpicii tablets discovered in Pompeii in the 1950s. The Sulpicii accepted deposits, lent money, acted as brokers, transferred funds between customers and changed money. Roman living standards were extraordinarily high compared to those of others at that time and this standard of living was only surpassed in the Industrial Revolution 1,500 years later. It was participation in a market economy that produced, bought, sold, transported, invested, lent, borrowed and innovated that gave the Romans a high standard of living. Romans were able to chase their own dreams. A market economy allowed this to happen. Slaves obtained an education and could buy their freedom. Freeing slaves was more common than in any other slave economy. Some 10% of Roman slaves were freed every five years whereas in the south of the US, it was 0.2%.

Natural formation of that dirty black sinful substance

Coal is of vegetable origin. It is organic. It is natural. Coal is a form of solar power because it's highly concentrated solidified photosynthetic energy. Burning of coal recycles carbon dioxide back to the atmosphere. It therefore should be revered by greens. It contains organic carbonaceous matter, inorganic material (minerals) and fluids such as coal seam gas (methane) and water. The carbon and hydrogen compounds in coal burn to produce heat, light, gas and a residual solid ash. Many coals contain fragments of fossilised wood, leaves, bark, fungus, pollen and spores and roots that go from coal into the underlying rocks.

In favourable settings, plant material dies, accumulates and decomposes, and can form coal. This decomposition is driven by fungus, bacteria and oxygen. If plant material is to accumulate and be preserved, it must be water saturated in a bog or swamp and covered with water to stop oxidation. The material that forms is peat. Bacterial activity almost ceases and the peat accumulates if it is not drained. There is a view that coal swamps formed in steamy Amazonian-type jungles. Nothing could be further from the truth.

Peat bogs form at high latitudes in cold climates where bacterial decay of accumulated vegetable material is very slow. The covering of peat with sediment, subsidence and compaction converts peat into brown coal and, with further pressure and temperature, converts brown coal into black coal and eventually anthracite. The process of coal formation takes place at depth and later geological processes uplift the coal seams to at or near the surface. For many technical and economic reasons, coal cannot normally be mined at depths greater than 300 metres.

This process of increasing the maturity from peat to brown coal to black coal results in the loss of water and an increase in carbon, coal density and the amount of energy that the coal yields. There are two major types of industrial coal, thermal coal and coking coal. Thermal or steaming coal is used to generate steam in coal-fired power stations to drive turbines for the generation of electricity. Coking coal is used for smelting to produce steel and metals. It is amazing that steam engines (e.g. the Newcomen engine) were invented in the 18th century, modified and improved by James Watt (1736-1819), further modified and are still used to provide the electricity so vital for living in the modern world. Just because some other technology is modern does not mean that it is the best available.

Underground mining of coal can produce dust. Coal dust, like many other dusts (e.g. wheat husk dust), can be explosive. During the mining of coal, waste rock is sometimes mined and coals might have internal seams of waste rock such as sandstone, shale or volcanic ash (tuff). Much of this material is removed by coal washing. Coal is mixed with water containing suspended magnetic iron oxide (magnetite) to create a dense liquid. Coal floats and waste rocks sink. The traces of magnetite stuck to the washed coal and waste rock are removed by magnets and reused.

Washed coal produces more heat, reduces transport costs and produces fewer waste products at smelters and coal-fired power stations. The best quality coals have low water, low ash (mineral and rock particles in coal) and low sulphur contents. Some coals in the Western part of the South Island of New Zealand have an extraordinarily low ash content (less than 1%) whereas most coals have an ash content of 10 to 20%. Some of the ash in coal cannot be removed by washing.

For more than 100 years, coking coal has been used to make steel, including the stainless steel for your teaspoon. Before coking coal can be used, a comprehensive suite of analyses needs to be performed. Measurements need to be taken to determine how quickly coal will wear out equipment. Tests of coking coal need to be undertaken to determine how much the coal will swell, when it will soften in the furnace, how strong the coal will be in a furnace, what gases will be emitted upon burning and when gas will be released from the coal.

Coal seam gas

During the process of conversion of peat to black coal by heat and pressure, gases are produced. The world's peat bogs, peaty tundra, swamps and soils leak methane gas into the atmosphere as do humans, bovines, termites and many other insects. This leakage process continues during conversion of peat to coal. In places, leaked methane (swamp gas) spontaneously ignites (with the help of phosphorus gases and microorganisms) making swamps appear dark evil spooky places at night with flickering lights, dancing phosphorescent glows and flames. For example, in Shakespeare's *The Tempest*, Caliban is frightened witless by lights in a swamp, probably from the combustion of methane (swamp gas). In more modern times, the odd dancing lights from swamp gas ignition have been interpreted as UFOs, aliens and will-of-the-wisps. Each to his own.

Gases produced by the conversion of peat to coal leak to the surface at variable rates, can be trapped in overlying rocks and are trapped in the fractures in coal. The main gas is methane although carbon monoxide, carbon dioxide, hydrogen sulphide and nitrogen are common.

Over the centuries, there have been some shocking multiple fatalities in underground coal mines from explosions of methane, carbon monoxide and dust. Mixtures of inert gases such as carbon dioxide and nitrogen have asphyxiated miners because of the lack of oxygen. Fall of the roof, walls and an inrush of water have also produced many fatalities. These explosions still take place in jurisdictions where the occupational health and safety laws are not as rigorous as in the Western world.

In order to avoid such gas explosions, modern mines are well ventilated and equipment that may produce a spark is not used. Electrical switching equipment is non-sparking and, where metal meets metal, zinc against zinc contacts are used. As iron can spark with striking silica-rich rocks, it is not used. Areas where tungsten carbide cutting teeth on mining equipment may spark are wet and ventilated. Matches, cigarette lighters and smoking underground are strictly prohibited.

Some coking coal contains a large amount of gas and, as an additional safety procedure, patterns of drill holes from the surface penetrate the coal seam to extract methane before mining as a safety precaution. For nearly a century, this coal seam gas has been bled from the coal and used at the mine site to generate electricity for the operation. Greens have only just discovered that there is such as thing as coal seam gas. However, those in the coal mining industry have known about it for centuries. So much for the "progressive" politics of the greens.

Uplift of coal-bearing sequences results in rock destressing due to unloading, the rocks fracture and gases migrate into fractures in coal. This is a process of natural fracking and is a common process at relatively shallow depths. Fracking is not generally used for coal seam gas extraction. With fracking at far greater depths for shale gas and oil, uplift has not naturally fracked the rocks so fractures are artificially produced by immense hydraulic pressures during the fracking process.

Because coal seams are generally enclosed by impermeable rocks, coal becomes a shallow reservoir for pressurised methane gas. Once the coal is penetrated by a drill hole, the gas drains from high pressure in the coal seam to low pressure where the drill hole has intersected the seam. The gas then naturally rises up the drill hole. The hole is cased, can be plugged at any time and flow can be adjusted with valves in surface facilities. The fingerprint of a coal seam gas drill hole surface facility is about the size of the average lounge room. There are millions of holes around the world each year that penetrate the water table and tried-and-proven procedures of casing holes to stop water table contamination are well established. Coal seam gas can be very cheaply extracted from seams that may be too deep or too poor in quality to mine. All that coal seam gas extraction does is accelerate the removal of gas that would naturally leak out of the coal over time anyway.

Underground coal seam mining does not remove all the coal. This is because pillars of coal need to be left for supporting the roof. In old mines, only about 30% of the coal was removed. In modern long wall mines, most of the coal is removed. Some weird and wonderful things happen in the underground world. A biochemical reaction between coal and water is driven by microorganisms and methane is produced. Openings in old coal mines are filled with methane. By drilling into these spaces (called goaf), methane can be extracted. What's more, the methane continues to be produced and the goaf operates like an underground gas storage tank. Again, drilling is through the water table using cased holes. It hardly needs to be said that entering an abandoned coal mine is the sort of action that leads to a Darwin Award.

There have been a number of corporate cowboys trying to rapaciously enter the coal seam gas industry for a quick profit, engaging in shocking practices in order to vent some of the non-flammable gases, such as hydrogen sulphide, to the surface and using cheap plastic casing rather than steel. Landholders have rightfully been incensed. These landholders have actually joined their mortal enemies, the greens, as anti-coal seam gas groups. It is not landholders and green groups that are running the cowboys out of town, it is the serious players in the industry and the regulators.

There is also great potential for offshore coal seam gas. Many coal basins are both onshore and offshore. One of the biggest basins is in the North Sea. There are an estimated three trillion tonnes of coal underneath the North Sea. This is the perfect setting for a massive coal seam gas operation that could power the UK.

Opposition by greens to coal seam gas extraction may be because of a lack of knowledge, because methane is a hydrocarbon gas or because anything associated with coal is off limits. I suspect that the real reason for opposition is that coal seam gas is a cheap source of energy and that coal seam gas-generated electricity bypasses the green control of wind and solar power. This opposition has nothing to do with the environment and everything to do with power over the average person by unelected minorities. Greens have never been elected to more than 50% of the seats in any parliament in the world so they enter coalitions with centre left parties such that the coalition can govern. In every case when there has been a coalition involving greens, unemployment has risen, productive industries have been destroyed, costs have risen, massive amounts of money has been wasted, freedoms have been lost and the government has flirted with totalitarianism. Again, the costs are borne by the average person, as per usual.

Sulphur

Most Northern Hemisphere coals have a high sulphur content. This is because the swamps from which the coal derived were close to the shore. Regular events of slight subsidence or sea level rise meant that peat swamps were sporadically inundated with sulphate-bearing seawater that was chemically reduced by decomposing plant material to sulphides. Some sulphur is also chemically bound onto organic compounds. Most large Northern Hemisphere deposits of coal are Carboniferous in age and the exhaust gases from burning this coal need to have the sulphur compounds scrubbed out. The geological time period named the Carboniferous is a good indication of what was happening at that time. Carbon dioxide was sequestered from a very carbon dioxide-rich atmosphere into plants. This modified plant material is now coal and, by burning coal, we are recycling sequestered carbon dioxide back into the atmosphere so that it can undergo another lap of natural sequestration.

In the mid 20th century, the burning of these high sulphur coals for household heating and cooking in the UK, Europe and USA created noxious fogs (pea soupers). Tens of thousands died from respiratory problems. This is exactly what is happening in China at present with small industry emitting particulates and sulphur gases creating dreadful smog and respiratory problems. In the Western world in the mid-20th century, this public health problem was solved by having large centralised coal-burning power stations that scrubbed particles and gases from exhaust gases. The domestic use of coal was banned. This was an environmental decision taken by responsible Western governments 60 years ago. No greens were involved. Heating and cooking at home was then by reticulated cheap electricity and not by burning sulphur-rich coals as previously. The smogs and respiratory problems disappeared. This shows that pollution can be stopped but only when communities become wealthier and apply simple engineering solutions to problems.

The first step to stop massive pollution in China from many small factories burning low quality coals has already been taken. Previously China would import low quality brown coal with a quality threshold of 3,491 k cal, 20% ash and 1% sulphur. Imports of these coals will be banned and so some 50 to 55 million tonnes of poor quality sulphur-rich coals will not be imported and burned. Some 96% of low quality coals have been imported from Indonesia. Although sulphur is scrubbed out of waste gases, even a small amount of sulphur oxides released into the air can cause acid rain and respiratory problems.

By contrast, Southern Hemisphere coals and those from India are slightly younger, formed in a deltaic environment up slope and further from the sea and accordingly have a lower sulphur content because of the lack of inundation by the sea. The large Southern Hemisphere coal deposits of India, South Africa, Australia, South America and Antarctica formed during very cold times. In places, coal seams are interlayered with debris left behind by retreating glaciers. This environment is also well represented in the modern swamps of the Northern Hemisphere where peat directly overlies glacial debris left behind by retreating glaciers at the start of the current interglacial (about 10,500 years ago).

Oil shale

Coals dominated by spores and pollen can be used to make petroleum. These are the oil shales (sapropelic or cannel coal). When lit with a match they burn, they float on water and were a source of petroleum in the 19th century and World War I. They are different from shales that hold crude oil and gas that are now being exploited by horizontal drilling followed by fracking.

These mined oil shales were crushed and heated in a retort, petroleum vapour was released and condensed to form various fractions of liquid hydrocarbons. Although a tonne of oil shale could release up to 100 litres of hydrocarbons, there were monstrous amounts of rock waste produced. This is exacerbated because after crushing, heating and distillation of petroleum, the residue has a volume 30% greater than the unmined oil shales and hence is too voluminous to fill underground or open pit space. Probably the biggest reserves of oil shales are in the Green River Formation of Wyoming, Utah and Colorado (USA).

Huge reserves of oil shales exist in many coal-bearing sedimentary basins as well as coal seam gas and deeper shale gas. Petroleum from oil shales is uneconomic unless the crude oil price stays above $US120 per barrel for a long time. Until then, oil shale sits on the reserves bench of the energy game and is a security against the closure of sea traffic or ultra-high crude oil prices.

Energy security

Few Western countries have energy security. The West is extremely exposed because greens object to an efficient mix of coal, gas, oil, nuclear and hydro electricity and have persuaded politicians to embrace feel-good inefficient ideological election-winning energy that can neither provide base load electricity nor the required energy density. To make matters worse, Western countries are preparing for speculated global warming by trying to reduce carbon dioxide emissions and are not preparing for an inevitable global cooling event. In World War II, the guns protecting Singapore were permanently mounted in one direction to repel an anticipated Japanese advance. The Japanese soldiers came from the other direction and Singapore fell. A lesson?

Most Western countries have low stockpiles of diesel fuel used for food transport from rural to city areas and we are always 10 days away from food shortages and civil unrest. Most Western countries do not have an energy disaster plan. Quick decisions in the absence of an energy disaster plan will need to be made in a time of crisis. However, Western politicians most commonly are lawyers and decisions take a long time. China has led the way with long-term energy, resources and food security. China is run by engineers and scientists.

Since 2005, the global oil production has started to decline and the oil price has risen about 15% *per annum.* Conventional oil production in the US is decreasing and unconventional oil from fracking will postpone the agony for about 30 years. Some 25 billion barrels of oil and 24 trillion cubic feet of gas have been discovered in the shale gas revolution. Maybe this has given enough breathing time for a clear-headed US energy policy? Oil consumption in the US has also fallen because people have driven their vehicles more slowly. Road deaths have been lowered and the distance covered by vehicles has decreased.

Coal is absolutely vital for energy security and preservation of oil reserves. The attack on coal by green activists is an attack on the energy security of the nation. In Weimar Germany in the 1920s, coal-to-liquid processes were invented and used in World War II because Germany was isolated from sea transport of oil. During embargoes, South Africa also used coal-to-liquid technology. The Bergius process uses powdered coal and hydrogen catalysed at high temperature and pressure and has a high yield per tonne of coal. The Fischer-Tropsch process has a lower yield, is not as complicated and burns coal in pure oxygen to make carbon monoxide and hydrogen. These are converted to liquid petroleum using a catalyst. The Fischer-Tropsch process can use poor quality and water-saturated coal.

These coal-to-liquids processes are viable when the crude oil price is above $US70 per barrel. Coal-to-liquids is the highest value use of coal resources, creates a very clean-burning fuel and provides the most convenient and lowest capital cost energy source for transport for consumers. Countries such as Australia have widespread coal resources of variable quality that occur in each state and huge volumes of hydrocarbons are currently transported great distances from refineries and ports. It is the obvious place for numerous regional coal-to-liquids plants.

At $US120 per barrel, all coal-fired power stations could be closed and replaced by nuclear power. Coking coal would still have to be mined. Coal-to-liquids could provide the required liquid hydrocarbons. For example, in the US about one billion tonnes of coal is burned each year for electricity. Known coal reserves and resources could keep the lights on for 250 years. Petroleum reserves are an order of magnitude lower. If the US wanted to replaced the seven million barrels of crude oil used daily, then four million tonnes of coal a day for coal-to-liquids would not only provide the required oil but would create employment and energy efficiency savings. The coal-to-liquids processes create a constant stream of carbon dioxide that could be pumped down exhausted oil wells to lower oil viscosity and recover some 34 billion barrels of residual oil.

Electric cars are a poor solution and cannot be considered as transport in any fuel security considerations. They are inefficient, the energy density of the lithium ion battery is an order of magnitude lower than diesel fuel and electric cars use 20% of their charge just to carry the battery. If electric cars use electricity generated from gas, they are not energy efficient. They may be energy efficient if electricity is from nuclear power stations. The most efficient transport at present is by burning liquid hydrocarbons in an internal combustion engine.

The only sensible solution to keep the Western world running for the next few hundred years, based on energy density, coal reserves, efficiency and declining petroleum reserves is to have nuclear power for electricity, coal-to-liquid for vehicle transport and dams for back up hydro electricity and food security. Coal seam gas, fracked gas and shale oil could be used for the chemicals industry and local energy sources. However, green activism is attempting to stop the use of nuclear power, dams, coal, gas and hydrocarbons. Do greens have a viable efficient alternative? No. Our children and grandchildren will pay dearly.

Save the whale, use coal, gas and oil

In the 18th and 19th centuries, we used whale oil for heating and lighting. This was replaced by fossil fuels in the late 19th century. What would you prefer for your heating and lighting? Energy from coal and oil or energy from whale oil? Australia is very critical of Japanese whaling in the Southern Ocean yet Australia only stopped whaling in late 1978. Other countries such as Norway and Iceland still slaughter whales.

Bob Brown is a former Green Party political leader in Australia. He is also the chairman of Sea Shepherd Australia. This is the same Bob Brown who fulminated when a grounded Chinese coal ship released some oil on the Great Barrier Reef:

> Studies of previous accidents shows damage to the Reef occurs through physical damage to the coral substructures and toxic pollution from marine anti-fouling paint, as well as impacts from oil spills ... Because of the sway the industry has over the government, the Great Barrier Reef has been turned into a coal highway ... The Greens are calling for a Royal Commission into how this situation could occur. Certainly, the coal industry should be held to account.

However, in early 2014, the anti-whaling ship *Sea Shepherd* pleaded guilty in the Cairns Magistrates Court (Australia) for polluting Great Barrier Reef waters with 500 litres of oil.

So, it's OK for an anti-whaling ship to pollute the Great Barrier Reef but not OK for a coal carrier to do the same thing. It seems that there is a pecking order of environmental concerns and that saving whales from Japanese fishing boats is a higher priority than dropping oil in Great Barrier Reef waters.

Apart from the *Cairns Post* and a few blogs, there has been no mainstream media mention of *Sea Shepherd* polluting the Great Barrier Reef. By contrast, the news outlets got themselves into a huge lather about a grounded Chinese ship releasing oil onto the Reef a few years earlier.

4

IN IRON WE TRUST

Three-quarters of the weight of stainless steel is composed of iron. It is a marvellous metal. It can be welded, cast, machined, forged, cold worked, tempered, hardened, annealed and drawn. It is very strong. No other material can do what iron does in today's world. It also rusts. Iron is very abundant. It is the fourth most abundant element in our planet, it is also the fourth most abundant element in the crust. Iron is everywhere. The most abundant element on Earth is oxygen that is bound in magnesium-iron-calcium silicate minerals in the crust and mantle. A huge amount of energy is required to release oxygen gas from minerals.

Oxygen gas is not a planetary gas on Earth, it derives from life. No other planet or moon in our Solar System has oxygen gas in the atmosphere and the race is on to see if there are traces of oxygen gas in the atmospheres of the 1,700 planets so far discovered outside our Solar System. For centuries we thought that there might be planets outside our Solar System associated with other stars, but it is only in the last 20 years that we started to find them. If the atmosphere of one of these planets contains oxygen gas, then there is a good chance that there is life on that planet.

Iron is in haemoglobin that is used to transport oxygen in our body. Without haemoglobin in our blood, breathing would be pointless. Iron is at the core of important enzymes without which we would die. Iron gives the red and green colour of soils. Iron-bearing dust particles in the air fall into the oceans, they fertilise algae and sudden bursts of oxygen are released into the atmosphere. Iron is dissolved in river, sea and ocean waters. Iron is released from submarine hot springs at mid ocean ridges and precipitates as sulphides and oxides. Iron is in more places than you can poke a stick at.

Iron, iron, everywhere

Although iron in minerals is very abundant, to make iron metal from iron-bearing rocks is not that easy. Metals are made by smelting, the process of converting rock to metal.

Earth magnetism

Most of the iron in planet Earth occurs in the core. The solid inner core is encased by the molten outer core. The core contains small amount of nickel, sulphur and other materials. During the first 100 million years of the Solar System, not only did planetary bodies form but the planets underwent gravitational settling. Although the Moon was sliced off the Earth by an impact very early in Earth's history, both the Earth and the Moon underwent later gravitational settling. Iron sank to form the core, this was covered by a cooled rocky mantle and bits of crustal scum that floated to the surface.

I know, I know, you don't believe me. Do the experiment. Order a Guinness at the front bar. Watch separation as the black stout sinks to the bottom (core) underneath the froth (mantle). This process takes time, as with core separation in early Earth. As with all science, experiments must be repeatable so this experiment should be done time and time again until you are satisfied.

Planet Earth still has a molten outer core enveloping a solid inner core. The cores of the Moon and Mars were each silly enough to freeze early in their history which had dire consequences, such as the loss of their magnetic fields and the stripping away of the Martian atmosphere and Martian oceans by solar wind.

Earth's swirling vortex of molten iron-nickel is 3,000 kilometres beneath the poles, it gives the Earth its magnetic field and is like a giant

2,400 kilometre-wide hurricane that moves about a quarter of a degree each year. This slight movement results in the wandering of the magnetic poles. Turbulence, fluctuation and eddies in the liquid iron-nickel outer core also cause fluctuations in the Earth's magnetic field.

Magnetic reversals

Both the direction and the strength of the magnetic field change. For example, in Roman times the Earth's magnetic field was half today's strength. The South Magnetic Pole was once in Tasmania.

At times, the Earth's magnetic field suddenly reverses. The North Pole becomes the South Pole and the north point of the compass points south. This has happened many times before. Over the last 50 million years there have been more than 100 magnetic reversals. It has happened when humans were on Earth. The good news is that life did not get wiped out. The poles will flip again. We don't know when. How would this affect us?

We're not sure whether a magnetic reversal would take days or tens of thousands of years. The only good record we have is the palaeomagnetic data from one lava flow that was cooling during the time the Earth's magnetic field flipped. It looks like the magnetic reversal took place over a 15-day period. During a reversal, we would be bombarded by more solar and cosmic radiation because the Earth's magnetic field operates as a shield to extraterrestrial particles. The Earth would become cloudier, cooler and there would be increased UV radiation. Telephones, television, radio, computers and satellites would not work. Planes could not fly. The electricity grid would probably collapse, there would be no accurate method of time keeping hence alternating current electricity could not be generated. Wind, solar and biomass energy would not work as there would be no alternating current electricity grid system. None of the comforts of modern life would operate. No electricity, no heating, no cooling, very little transport and so back to mother Earth.

Many in the cities would go back to mother Earth. In graves. There could be a rapid mass collapse of Western society and many would die, especially in the cities. Those in rural and undeveloped areas would be relatively unaffected. We've already had a taste of this scenario when the Earth gets bombarded during solar flaring. There would be no Twitter. No Facebook. No blogs. No cell phones. During a magnetic reversal, we would probably actually sit around and talk in the dark, probably around a fire that would serve for both cooking and heating. It would be a greens' paradise.

The wandering of the magnetic poles and reversals of the Earth's magnetic field have been used to reconstruct continental drifting. It is strange but true. The continents drift around, they still do and move up to 20 centimetres per year. Sometimes a continent will drift across a pole and there will be a glaciation.

Every time lava erupts, the magnetic minerals preserve the Earth's magnetic field at the time the lava solidifies. Reddish-brown magnetic iron minerals in soils also preserve the Earth's magnetic field. If the age of the lava or soil can be determined by independent means, then the position of the old magnetic poles at that time can be located. These pointers to the poles are later dragged across the surface of the Earth on drifting continents and a reconstruction can show where the continents were at a particular time, the position of the poles at that time and the rate of continental drift. Over a short geological time span such as 20 million years, continents can move 1,000 kilometres.

Life on Earth is protected from intense radiation by the magnetic fields of the Sun and the Earth. The Sun's magnetic field reduces incoming cosmic radiation whereas the Earth's magnetic field reduces both cosmic and solar radiation. We humans are creatures that live with almost constant radiation and a wide spectrum of electromagnetic energy. When the Earth's outer core eventually freezes, the atmosphere and oceans will be blasted into space by solar wind. Life on Earth will be fried. It happened on Mars. It will happen on Earth, also on a Thursday, in about 4,500 million years time. Put a mark in red, "END OF THE WORLD", on the kitchen calendar.

Because the Moon is so small, its core froze a long time ago. The Moon had a magnetic field about 3,600 million years ago. It probably had a small thin atmosphere that was also blasted away by solar wind. It now has no magnetic field. Because the Moon's core is solid, there is no transfer of heat from the core to the lunar crust. The lack of moonquakes supports this suggestion. The Moon is essentially dead because it lost its magnetic field. Many people I have met are essentially dead because they also have no magnetism.

Getting into a lather

Iron is abundant on Earth and in space. The fact that we have iron in our Solar System tells us that the Sun and its planets are 4,500-million old recycled stardust that initially formed 13.8 billion years ago. Our planet has regular visitors from space. Large visitors (asteroids) impacted with our planet early in Earth history and, thankfully, the chances of a massive asteroidal impact have decreased over the last few thousand million years.

We need to get asteroidal impacts into perspective. There is nothing we can do about an asteroid larger than 10 kilometres in diameter striking Earth. These impacts cause local and global extinctions and short-lived climate changes.

Some 99.99% of all life that has ever existed on planet Earth is now extinct and impacts are just one of the many triggers for extinction. All life on Earth is eventually heading for extinction and the vacated ecosystems are filled by other species that also will enjoy a short time on Earth.

Our planet has enjoyed regular asteroidal impacts, supervolcanoes, tsunamis, sea level rises and falls, six major ice ages (in which there were hundreds of glacials and interglacials), releases of poisonous gas, sudden events of increased solar and cosmic radiation, exploding atmosphere, trans-species disease pandemics, rapid habitat changes and appearances and disappearances of short-lived dominant species.

The Earth is OK, it will keep on doing what it has always done whether we humans exist or not. It has survived everything thrown at it in the past and these past events have been far greater than anything humans have created. The area of forest lost by land clearing for crops is far less than the area of devegetation that occurs during glaciation. The slight falls and rises in temperature that are currently measured are well within the historical and geological variation of temperature changes on Earth. The atmospheric carbon dioxide content is very low. If it halved then there would be no life on Earth. If it doubled, life on Earth would thrive as it did in the past.

The Australian territory absorbs up to 20 times the amount of carbon dioxide we emit. We are duty bound to release carbon dioxide into the atmosphere to keep vegetation growing and to maintain life on Earth. That is real sustainability. In the past, it has been emissions of sulphurous gases that have shortened human lives and killed other life. We can now scrub these gases out of our emissions from burning coal and oil and have reduced industrial emissions of harmful sulphurous gases.

If the greens really want to get into a lather about something, then it should be about the next inevitable glaciation. There have been numerous international conferences attempting to bring in protocols to prevent global warming. Why not international conferences to prevent global cooling? Internationally agreed protocols developed after numerous conferences and junkets to pleasant parts of the planet could rule that the Earth's axis stop changing and fluctuations in solar radiation cease forthwith. This would have the same effect as the unscientific and unenforceable protocols to stop global warming. If greens knew something about the planet they wished to save, then there would be preparations for the next inevitable minor disaster on our dynamic planet. Once the planet stops being dynamic, it will be dead. Like Mars.

Impacts, asteroids and meteorites

Orbiting between Mars and Jupiter are asteroids. They are thought to be the shattered remnants of small bodies formed within the Sun's solar nebula that never accreted enough to become a planet. These are leftover planetary building blocks that have a number of different orbits. There are probably about 25 million asteroids with a diameter greater than 100 metres. Most fragments were pulled into Jupiter by its very strong gravitational field and Earth gets sporadically bombarded by the escapees. As of February 2014, some 10,576 near Earth objects have been discovered. Of those, 97 are near-Earth comets and 10,480 are near-Earth asteroids. Approximately 1,450 near-Earth objects are potentially hazardous.

Planet Earth has been constantly peppered with meteorites and comets. The history of meteorite impacting can be pieced together by looking at craters, at shock features that occur deep in the Earth and at ejecta horizons. With some very large impacts, there is a coincidence of a mass extinction with an impact. There are three major types of meteorites: irons, stony irons and stony. Stony meteorites are the most common. Stony irons are material that was from the mantle-core boundary of a fragmented proto planet that had not completed core-mantle separation. The irons represent the core of the proto planet. Iron meteorites are an iron-nickel alloy.

Although more irons are found than stony irons (because they are far more visible), about 7% of recorded meteorite falls are irons. Deserts and ice sheets are the best place to find meteorites. For thousands of years, the Inuit people collected iron meteorites from ice sheets and fashioned them in to weapons and tools because the iron-nickel alloy is very hard and is relatively resistant to corrosion. The largest known iron meteorite is 60 tonnes in weight and lies in a field at Hoba in Namibia. There was once an attempt to mine iron meteorite fragments from the Meteor Crater, Arizona. It failed.

Comets

It is harder to piece together the blow-by-blow history of comet impacting because comets and some meteorites explode in the atmosphere and leave very little extraterrestrial material on Earth. There are consistent patterns in ancient writings. Over the last 5,000 years, it appears that comets have been responsible for famines and mass social disruption. It is no wonder that comets were once interpreted as portents and signs of difficult times in the future.

For example, a period of cometary impacting from 2,354 to 2,345 BC was one of untold misery, famine, destruction of cities and populations, earthquakes, tsunamis, volcanic eruptions and poisoning by gases released from the oceans. The falling of stones from the heavens weakened Egypt and may have led to the Exodus. Cometary activity was high about 1,600 BC as recorded in the Bible and Chinese writings. Impacting between 1,159 and 1,141 BC may have triggered the famine in the Biblical account of King David's reign. Ephorus observed the fragmentation of a comet in 372 BC.

Nomadic hunters and gatherers were better adapted to sudden global changes resulting from magnetic field changes or impacting than an agricultural civilisation that depended upon optimal sunshine, temperature and rain. Urban populations today are even more fragile because they depend not only upon agriculture and transport but a complex infrastructure vulnerable to tsunamis, earthquakes, fire, flood and politics.

If the greens disappear into the forests now and live as hunter-gatherers, they would be poised to survive the next inevitable asteroid or cometary impact and we carbon dioxide-emitting sinners living in the modern world would die. This way the greens will end up running the world without having to resort to deception. It is therefore the greens' moral duty to disappear into the forests. Now. And don't come back.

Crustal iron

The mantle of the Earth contains iron locked in silicate, oxide, sulphide and iron alloy minerals. The iron we use comes from the crust of the Earth. Although the crust contains an abundance of iron in silicates, the forces that hold iron, oxygen and silicon together in common rock-forming minerals are so great that a huge amount of energy is required to extract iron from common surface minerals. A very small amount of the iron we use derives from roasting iron sulphide, capturing the sulphurous gases for manufacture of sulphuric acid and using the residual iron oxide for iron manufacture. Metallic iron is very rare on the surface of the Earth because iron oxidises (rusts) so easily. It occurs in iron meteorites and in some chemically unusual oxygen-poor volcanic rocks (such as the basalts of Kassal in Germany).

The iron we use for steel is from the crust of the Earth and is contained in oxide, hydroxide and carbonate minerals. The extraction of oxygen from iron oxides and hydroxides is done during smelting. Iron carbonate ores first need to be heated to remove the carbon dioxide and produce iron oxide and then to be smelted to remove the oxygen. These processes for converting rock into iron were invented in the dim distant past.

OK, we have been distracted enough and need to get back to your stainless steel teaspoon soon otherwise you'll never get to use it.

Finding iron ore

Ore is a mineral or aggregate of minerals from which a valuable constituent, especially a metal, can be profitably mined or extracted.

In former times, if an iron ore deposit occurred at the surface, experiments were undertaken to determine if it could be smelted into iron using local materials. Mineral exploration was an extension of geographic exploration and empire building and iron ore mining continued in areas where there had been a long history of previous mining. There was no science to exploration. Mineral exploration today uses leading edge mathematics, chemistry, physics and geology.

Poisonous oxygen

Although iron ores occur in rocks of many ages, the best formations to host iron ores are between 2,600 and 2,400 million years old. This was the time on Earth just after the first major ice age. It was the time when the continents started to thicken up and the time when bacterial life started to exhale oxygen gas. As the oxygen gas built up in the atmosphere, soils changed from green reduced ones to red oxidised ones, prokaryotic bacteria started to die from oxygen poisoning, eukaryotic life evolved and started to pump out even more oxygen and oxygen started to dissolve in the ocean. Prokaryotic life still exists as refugees from oxygen in your stomach, reduced rocks, bogs and swamps.

Once there was sufficient oxygen dissolved in the oceans, the dissolved reduced iron in the oceans started to oxidise and was precipitated as insoluble oxidised iron oxide on the sea floor together with large quantities of silica. This was a time in geological history when the oceans suddenly rusted. The precipitate was a finely laminated sludge composed of layers of opaline silica and iron oxide ooze (hæmatite, Fe_2O_3) that was later hardened into a rock called banded iron formation. These banded iron formations contain 15 to 25 % iron, not enough for an economic iron ore deposit.

Iron ores and climate change

However, nature did the mineral processing upgrade for us. When banded iron formations are exposed to a long history of tropical weathering, silica is removed in solution and the iron oxide remains. Iron oxide remnants cap hills (e.g. Mount Tom Price, Mount Newman) and, as the peaks of the hills collapse downslope, they form new iron ore deposits. The fragments from erosion of bedded deposits can be deposited in landslides and ancient river channels in economic concentrations where natural traps existed. These are referred to as detrital iron deposits. Economic deposits of iron ore can also form through the deposition of tiny eroded grains of hæmatite in old river channels. These are called channel iron deposits and appear to be unique to Western Australia. Channel iron deposits can often have a very high proportion of water compared to high quality lump ore. Today a number of companies classify low-grade, highly contaminated iron-rich material as detrital iron ore on the basis that it can be beneficiated to produce an economic product.

Channel iron ore deposits can be reworked by nature. The grains in detrital deposits are deposited by erosion and the iron in ground water accretes around the grains to form pea-sized grains known as pisolites. The pisolites are in turn cemented by further deposition of iron. Most channel iron ore mined in Western Australia is direct shipping ore and because of the nature of the iron (hæmatite with a goethite matrix) these deposits tend to be lower in iron content.

Most of the iron oxide remaining after tropical weathering contains more than 50% iron and the ores mined generally contain more than 60% iron. The Pilbara iron ore deposits in Western Australia enjoyed more than 100 million years of tropical upgrading before the change to the current cooler climate 5 million years ago. The reason why we have rich iron ores in Australia, Brazil and West Africa is because of climate change.

Real climate scientists

But that's not new for geologists, we have been studying climate change for hundreds of years whereas those who call themselves climate "scientists" have only been studying climate for a few years and have regressed into climate activism. Many of these so-called climate "scientists" have no rigorous background in science and are very narrow in their field of expertise. Furthermore, if geologists get it wrong, they may lose their jobs. If a climate "scientist" gets it wrong, the climate "scientist" just applies for another research grant to try to show that we humans are hurtling towards the end of the world and it's all our fault.

Climate change has recently been discovered by the media and the general public and a whole new army of lickspittles now live off the climate industry. These people could not get a job in private industry or could not create a small business. More than a century ago, climate change was taught in university elementary geology courses.

In my book collection is a leather-bound set of hand-written geology lecture notes from 1890. The student was a J.G. Blackmore at The University of Adelaide. More than half the first year geology course by Professor Tate was on climate change. This book was given to me as a thank you present for a week-long specialist course given in Freiberg (Saxony) and originally came from Johannesburg where Blackmore must have worked after graduation.

The notes show that more than 125 years ago, geologists were well aware that climate changes, glaciers advance and retreat, sea level rises and falls, land rises and falls, coral atolls grow when there is a relative sea level rise, the planet is dynamic and the past is the key to the present. Some of the notes are also on how to create a geological map.

Exploration

In the modern world, the process of exploration involves creating a geological map that shows all the rock types, folds, faults and surface topographic and cultural features such as hills, watercourses, roads and buildings. These maps are created by an integration of field work on foot and from aerial and satellite surveys that measure the Earth's magnetic field, gravity, electrical properties and reflectance of energy.

Many iron ores are magnetic and these can be detected from the air. Iron ores are generally denser than the surrounding rocks, gravity is slightly higher over dense rocks and this can be measured. If you stand on an iron ore deposit, you weigh more than if you stand on many other rock types. Some iron ores conduct an electric current, other rock types are good resistors and some rocks preferentially absorb infra red and ultra violet radiation. All of this can be measured from the air. Once a suitable area has been located, ground work comprising more detailed measurement, mapping and drilling is undertaken.

Initial drilling is triple tube diamond core drilling. The core is a precious sample from the unseen depths that validates or refines the three-dimensional picture deduced from geological and geophysical mapping. Once the three dimensional geological picture has been validated, percussion drilling is used for further validation and more detail. Percussion drilling is relatively cheap and fast. A rotating hammer breaks rocks into powder and chips that are blasted up inside the drill rods by compressed air. These samples are collected every metre and a three dimensional picture of the rock type, rock hardness and rock chemistry is produced. Thousands of holes are drilled and very sophisticated statistical techniques are used to determine what is ore and what is waste, what ore types should be blended and what shapes should be designed for extracting the ore. These three dimensional pictures are windows into the probability of ores of a specific chemistry at depth and this is commonly integrated with a financial analysis for funding a new mine.

Once the explorationists have discovered a new iron ore resource, the data and three-dimensional pictures are handed over to a large team of mine geologists, mining engineers, mechanical engineers, civil engineers, electrical engineers, mineral processors, environmental engineers, hydrologists and economists to design a mine. The bankable feasibility study for a large iron ore mine costs many millions of dollars and takes years to complete. Normally there are forward projections made of the iron ore price and markets and very often much of the ore of a specific quality is contracted to be sold before mine construction even starts. This is a method of lowering risk. In Australia, not only do mines have to be built but towns, railways and ports may need to be built by the operator as mining is often in areas with little infrastructure.

Because of the volumes and costs involved in an extremely competitive market, modern iron ore mines are large and no small deposits, such as those mined in the past, can now be profitably mined.

Mining of iron ore

Pre-mining

A large iron ore mine is a sight to behold. Before mining the reserves (accurate estimation of ore in the ground) and resources (less accurate estimation of ore in the ground), the vegetation is sampled, seeds are collected and the soils are stripped and stockpiled for later revegetation. The first great environmentalists in Australia were the mining companies that built the regeneration reserve around the zinc-lead-silver mining town of Broken Hill (NSW). They undertook research on local climate, soils and vegetation before constructing the regeneration reserve.

The Broken Hill regeneration reserve stopped shifting dune sands, created a green belt, increased rainfall, decreased summer temperature and increased winter temperature. This was a great success. And when was it done? In the 1930s, before any green was born. For 50 years after the building of the regeneration area at Broken Hill, the local mining companies would give away seedlings to employees and towns people as part of the greening of Broken Hill. The major iron ore mining companies in the Pilbara of Western Australia, Rio Tinto and BHPBilliton, started life in Broken Hill and use similar environmental practices in the Pilbara to those initiated in Broken Hill 80 years earlier.

Blasting

Before mining, the calculated ore blocks are systematically drilled and explosives are placed in those drill holes for blasting of the ore. The chips and dust from the drill holes are chemically analysed to validate the ore block model and make last minute slight changes to the mine plan. After blasting, the boys and girls move in with their big toys. In one Western Australian mine, Hitachi EX5000 excavators are used, they weigh 53.3 tonnes and carry 27 to 29 tonnes in the bucket.

Some excavators have an X-ray analyser at the leading edge to chemically analyse the ore in the face of the open pit before the excavator picks it up. The analysis is transmitted to Perth and an almost instantaneous answer comes back from the operations centre to show whether the bucket load is waste, low grade ore or ore. Bucket loads are dropped into haul trucks.

Haulage

At this same Western Australian mine, the Komatsu 830E haul trucks are driverless. Before the start of the shift, the truck is programmed for the day's work and does not have to stop for morning tea, lunch or afternoon tea. The truck does not go on strike, go slow, get tired or stop for a fag and works non-stop until it needs fuel. These trucks are monstrous: 16 cylinder 60 litre diesel engines that deliver 1,865 brake horse power, travel at a maximum of 64 kilometres an hour, have a payload of 147 cubic metres (230 tonnes) and fully laden weigh 385 tonnes. Humans in other vehicles must keep 50 metres distant from these monsters and there are many safety photographs around mine sites showing what happens when a 385 tonne haul truck runs over a four wheel drive mine service vehicle. All that is left is a plate of steel.

Haul trucks travel under an arm that measures the load chemistry by X-ray fluorescence. The decision is then made at the operations centre in Perth whether the load goes to the run-of-mine pad (ROM), low-grade stockpile or waste dump. The exact chemistry of loads of ore dumped on the ROM pad is known, ore is mixed on the ROM pad and then crushed, sized, sieved and stockpiled. Movement from the ROM pad to the stockpile is by conveyor belts. At many stages in the process, the ore chemistry is measured.

Nuts and bolts

The full story is much more complex than the above outline might suggest. The mine operators need to know the total material movement, if not then a mine can go broke very quickly by shifting too much waste rock and not enough ore. The blasted tonnes of ore have to be reconciled with the tonnage mined. The tonnage of dry ore needs to be reconciled with the tonnage of wet ore. Machines have to operate efficiently and truck utilisation, drill utilisation, dig fleet utilisation, auxiliary fleet utilisation, fleet performance and scheduled and unplanned machine maintenance all need to be mathematically analysed.

There have to be enough machines such that most are working for 85% of the time and the rest of the time are undergoing scheduled preventative maintenance or unscheduled repairs. Workshops are massive and, because mining is in remote areas, operators fly parts in from all around the world and must be able to do everything from changing a tyre to rebuilding a haul truck from the ground up.

All vehicle communication systems must operate continuously, all vehicles must have fire prevention and first aid equipment and all vehicles must have efficient brakes, tyres and safety systems. Mine scheduling makes sure that there are no traffic jams and that different types of ore can be mined concurrently in different places for later blending. Equipment scheduling makes sure that an excavator is not waiting with a bucket of ore until a haul truck finally shows up. Slight changes in maintenance, planning or scheduling can have a profound effect on mine performance.

Geologists and mine planners work together to make sure that rocks associated with iron ore that may contain the carcinogenic blue asbestos (crocidolite) are neither blasted nor exposed nor occur as contaminants in ore that would undergo crushing, sieving, mixing and transport. Even though some areas might be rich in ore, they are not mined for mineralogical reasons (e.g. if there is too much blue asbestos or if there is too much iron sulphide). Iron sulphide oxidises to acid runoff waters and the last thing a smelter wants is sulphide minerals in the feed. Upon oxidation in the smelter, this produces sulphurous gases that create choking pollution and acid rain.

Some mines need draining and much of this water is used for dust suppression. The effect of mining on the water table is constantly monitored, an excess water strategy needs to be devised and plans need to be put in place for the inevitable cyclone. When an open pit is wet, large equipment cannot be used safely and train lines may be washed out.

Concurrent with mining, land and cultural heritage matters need to be managed. Mining takes place with a mine closure plan in mind. Most mine closures result in removal of all surface facilities and revegetation. If a mine is in a very sensitive area, then an open pit may have to be filled in after mining. This is very rare. Some areas of ore are not mined because they are close to areas of indigenous heritage and trees, landforms and rocks may be fenced off as part of cultural heritage preservation. This happens at most mines.

Furthermore, the mining industry in Australia is the biggest employer and trainer of indigenous people, especially in remote areas. What do the greens do for indigenous training and employment? Nothing. The workforce is highly skilled and undergoes constant upgrading of skills. Training is a big part of the job. Safety is paramount, safety training is constant and preparation for emergencies is part of this training.

Processing

In the processing plant, there are regular scheduled shutdowns for maintenance and the occasional shutdown when there is a catastrophic failure of equipment. Equipment in the processing plant is abraded, shocked and shaken and at times metal fails or needs replacing. The work index of the ore is regularly measured in order to get an indication of how long steel can be used before it is ground away.

Again, mines need a stockpile of blended ore ready to load on trains for the port such that when the processing plant is undergoing maintenance, product can still be shifted. Frequent cyclones are such that mining and rail transport stops and stockpiles are used to keep the wheels of industry spinning. Stockpile management is vital and the mine needs to manage buffers of broken stock, low-grade stock, pre-processing plant stock, post-processing plant stock and port stock. Management has contingencies factored in and is able to tune the rate of mining with the rate of stockpile depletion.

The stockpiled ore is separated into lump and fine ore and again stockpiled for loading onto trains. These trains are also a sight to behold. A number of 130,000 tonne trains leave the mine site each day. Some of these trains are driverless, they pass each other on sidings on the trip to ports that can be hundreds of kilometres away. If caught at a railway crossing, turn off the engine and make a cup of tea. It will take that long for a train many kilometres in length to pass. At the port, the train is automatically unloaded. The vibration on the long train trip grinds some lumps into fines, the ores again need to be sieved and again blending takes place to create a product that is of constant quality. Smelters around the world need feedstock of constant size, mineralogy and chemistry for decades in order to be efficient.

The end result of all this is to produce lumps 6.3 to 31.5 mm in size such that there can be an upward flow of gases in the smelter. Fines are sintered or pelletised into lumps. Lump product sells for more than fines. With both lumps and fines, the aim is produce a product that has about 61.5% iron, 4.3% moisture, 3.8% silicon, 1.6% aluminium, 0.18% sulphur and 0.05% phosphorus. Each major iron ore mining company markets its own unique blend of iron ore. For example, Rio markets its Pilbara Blend that is a mixture of the Brockman Iron Formation and the Marra Mamba Iron Formation.

Popular myths

Some tropically weathered banded iron formations are often of sufficient quality to be classed as direct shipping ore, which means they need negligible beneficiation, often just crushing and screening, before shipping to the customer. These are the jewel in the iron ore miner's crown because some large capital costs are then avoided. One of the largest hæmatite deposits in the world (and the biggest mine, 5 kilometres by 0.5 kilometre) is Mount Whaleback, operated by BHP Billiton at Newman in Western Australia. It had an original resource of 1.7 billion tonnes of high quality iron ore at a grade of 64% iron.

In Brazil, the Carajãs mining district contains multiple hæmatite deposits with an initial total resource of more than 17 billion tonnes with grades of over 64% iron. We certainly won't be running out of iron ore during this iron age that has been going for thousands of years and will go well into geological time. Speculation about the exhaustion of mineral resources on Earth is just scaremongering.

There is a view that one often reads in *Letters to The Editor* that mining involves just digging up material and exporting the bulk product. If only it were so easy. Such views indicate a total lack of basic knowledge by the chattering class common taters. Not only does Australia export minerals but it also exports mining skills, software and machinery.

Maybe the summary above shows that a relatively easy mining operation such as open pit iron ore mining is a highly sophisticated scientific, mathematical, engineering and financial exercise with a constant eye to safety, innovation and improvement of performance. The risks are huge, investments must be made over decades and all people in Australia are beneficiaries of mining. Governments cannot just change the laws and regulations to suit short-term goals or green whims because the exploration, construction, operation and mine rehabilitation of a large iron ore mine can take a century.

Small iron ore deposits

Iron ores in former times were pretty dreadful. They were low grade, tonnages mined were not large, there were many impurities and the ores were highly variable. This made smelting to produce iron a challenging process. These ores serviced local foundries and smelters. Most were mined from shallow open cuts and others were mined from small dangerous underground mines. Both mining methods had no geological or engineering input. Most of these ores formed from springs or from precipitation in bogs and shallow marine settings. The 20th century required larger tonnages of iron ores that had a constant chemistry throughout the life of the mine and the small iron ore mines were replaced by large mines.

Springs entering bogs carry iron in solution that is oxidised by air and bacteria (*Acidithiobacillus ferrooxidans*) when it enters the bog. Goethite (iron oxy hydroxide; FeO[OH]) precipitates. These bogs are in areas that were recently covered by ice and are common in Europe, Russia, Canada and USA. These again tell us that the planet is dynamic, that huge ice sheets covered much of the planet until about 10,500 years ago, that climate change is normal and that the temperature, landscape and sea level rate of change in the recent past were far greater than changes occurring today, whatever their origin.

Bog iron ores contain about 40 to 45% iron with high quantities of silicon, aluminium, calcium and magnesium minerals, a high phosphorus and sulphur content and about 10 to 15% water. They are a poor ore with a low iron content and a high proportion of impurities. In the past, dark brown pebbles of bog iron ore were mined from shallow waters along the shore of swamps and bogs. After a few years, a new layer of bog iron ore formed.

They were mainly mined in pre-Roman times and in the Viking era. When better iron oxide ores were found in the 16th century, the mining of bog iron ores gradually declined. Mining of bog iron ores stopped in the early 20th century because bigger, better and more economic deposits displaced them as a source of iron ore. Bogs that contain iron ore have an iridescent oily coating as an iron oxide slick on the surface of bog water.

Siderite (iron carbonate; $FeCO_3$) iron ores form in a shallow marine setting in much the same way as bog iron ore. An iron oxide/oxyhydroxide cap formed on siderite layers after the end of the last glaciation. At Eisenerz (Styria, Austria), siderite ore has been mined since the 12th century. The deposit originally had an iron/oxyhydroxide cap that formed after the glaciated peaks were exposed to interglacial warm wet conditions some 8,000 years ago. Geologists have known for hundreds of years that these caps formed in post-glacial times from natural climate change and that some geological processes can be extraordinarily rapid.

The Donnawitz Valley downslope from the Eisenerz mines was the centre of the iron trade, scores of smelters operated in the Middle Ages and a number of ingenious relics of alpine transport and smelting still remain. Once the iron oxide/oxyhydroxide ores had been exploited, the iron carbonate ores were mined at Eisenerz. Similar iron carbonate ores were mined in USA, Germany and England. In the UK, some 75 separate layers of siderite were mined during the Industrial Revolution, these siderite ores provided about 80% of the iron produced at that time.

In other marine settings just beyond the strong currents and wave base, iron carbonate-iron silicate-iron oxide precipitates form. They are composed of millimetre-sized balls of these iron minerals. These have been mined as a poor quality iron ore in USA, Canada, France, Germany and the UK. Upon weathering in interglacial warm wet conditions, residual iron oxides/oxyhydroxides form and are richer in iron than the original carbonate-silicate deposits.

Major iron ore deposits

Types

There are two common iron oxide minerals, hæmatite (Fe_2O_3) and magnetite (Fe_3O_4). Both are abundant. Hæmatite as crystals is black and flakey and when it is massive, it is red, brown or black. When hæmatite is powdered, the colour is red. In many parts of the world it was called bloodstone and massive finely crystalline black hæmatite is still used as a semi-precious stone. For thousands of years, powdered hæmatite has been used in cosmetics and paints. It still is. Hæmatite contains 70% iron. The hæmatite deposits are the major source of iron ore described above. A lower tonnage of iron ore is from magnetite ores.

Magnetite iron ores

Magnetite has also been known for thousands of years. Magnetite contains 77% iron. There are many large operating magnetite mines in the world (e.g. Kiruna, Sweden). Magnetite is called magnetite for a very good reason. Needles of magnetite were used as compasses and were called *magnes*, *magnetis*, *heraclion* and *sideritis* by the ancients. Most commonly the needle form of magnetite was called lodestone. Many Greek writers such as Hippocrates and Aristotle described lodestone at length. Theophrastus, Dioscorides and Pliny also describe lodestone.

In many parts of the world, the heating of soils by molten rock forms a rock composed of magnetite and corundum (Al_2O_3). Pure gem corundum is sapphire and ruby. The mixture of hard magnetite and the even harder corundum is called emery and has been used for thousands of years for polishing gemstones. It still is. Emery is still mined in the Menderes Mountains of Turkey.

Despite a lower iron content, hæmatite is the preferred iron ore source, magnetite also forms as an intermediate product during the smelting of hæmatite and dedicated smelters must be used for the smelting of magnetite. Hæmatite is easier and less expensive to smelt than magnetite.

In ancient rocks, layers of quartz-magnetite rocks are common. These were originally sea floor hot spring precipitates. The iron content in quartz-magnetite rocks is about 20 to 35% iron. Most of these quartz-magnetite rocks are not viable. They are not ore. Quartz-magnetite rocks need to be beneficiated to an intermediate product. Typically, the magnetite and silica must be able to be separated after highly expensive fine grinding to no more than 30 to 45 microns (micron = micrometre = one millionth of a metre). The magnetite is far denser than quartz and can be separated by gravity methods or by using electromagnets (because it is magnetic). The magnetite slurry must then be dried and filtered. It is then typically rolled into balls or pellets and roasted. The concentrate should grade in excess of 63% iron and have a low phosphorus, aluminium, silica, sulphur and titanium content.

These quartz-magnetite rocks are folded, corrugated and often bent double. Even in areas covered by glacial debris such as Quebec and Ontario (Canada), Sweden, Finland and Russia, such deposits can be found by measuring the Earth's magnetism from aeroplanes. Many quartz-magnetite rocks form from exposing quartz-hæmatite rocks to heat and pressure. In many places, quartz-hæmatite masses occur within quartz-magnetite layers and can be preferentially mined (e.g. Lake Superior, USA).

If a quartz-magnetite rock is exposed to the air for a long period of time in warm wet conditions, the quartz will be leached into ground waters and magnetite will oxidise to hæmatite. This is often mined (e.g. Brazil). In others places, the hæmatite weathering cap has been removed by moving ice sheets during the last glaciation (e.g. Canada).

Size

Size counts. That's why there are now no small iron ore mines in the world. With modern mining, the greater the tonnage shifted, the lower the costs per tonne. Banded iron formations are the most important source of iron ore today. They occur on all continents and the total contained iron content can be extremely large. They are dominantly quartz-hæmatite rocks.

In Australia, the Hamersley Range in Western Australia is thought to have initially contained 100 trillion tonnes of iron. This is probably less than 5% of the world's total iron in banded iron formations. A trillion is not a large number in today's inflated world, especially in terms of government debt. At current production rates, 100 trillion tonnes would be enough to last the world for at least a few thousand years. The world is not short of iron or iron ore. Within that 100 trillion tonnes is high-grade ore (>60% iron) and low-grade ore (55-60%) which is currently stockpiled.

Ore quality

The most important factors in determining the economic value of bedded iron ore are the physical and chemical characteristics. The ore must be able to be crushed without producing an undue amount of fines, which impede gravitational draw down in a smelter. Transport and multiple handling degrade a small proportion of lumpy ore into fines. Fines can be easily dealt with but at a cost. Generally, hæmatite ore needs a minimum grade of 55% and realistically greater than 62%. The lower the iron content, the higher the contaminants. However, lower grade bedded iron ore can often readily be beneficiated to direct shipping ore at low cost. Ideally iron ore contains only iron and oxygen. This is never the case.

The most significant contaminants are silica, alumina, manganese, titanium, sulphur and phosphorus but there are many other elements that are deleterious to the value of iron ore. Phosphorus is the bad boy. There are about 8 to 10 billion tonnes of high phosphorous (more than 0.10% by weight) iron ore in Western Australia that no one wants. Phosphorus today is impossible to remove from banded iron formation ore. If the greens want to save the world, then they should invent a process for removing phosphorus from iron ore and making a fertiliser to help feed the poor in the Third World.

In iron ore, silica and alumina should be below around 5% and 2% respectively. Anything much higher would generally not be considered ore. Both can be removed by beneficiation and fluxing in the smelter, but this can be a significant expense. Another important variable is that maximum loss on ignition must be 7 to 10%. Loss on ignition is defined as the amount of water that vaporises at a temperature of 1,000°C and this includes chemically bound water as well as moisture. For many solid particles, there needs to be a certain amount of water for tipping and pouring crushed material but not enough water to create liquefaction to a slurry from vibrations during ship transport. Furthermore, it is not a good economic idea to transport too much water in iron ore and to vaporise water in smelters (which costs energy).

Making iron

Ancient smelting

Fluxes lower the melting temperature of materials and hence save energy. The modern world uses fluxes for soldering, welding, glass making and smelting. In the dim distant past, iron ores were used as a flux for smelting copper ores and the red iron oxide hæmatite has been used in copper smelting to separate gold from copper. Iron ores are still used as a flux in copper, zinc and lead smelters. The use of iron ores as a flux for copper smelting probably led to the discovery of iron metal.

We don't know when iron smelting was first discovered. Many places in the Bible have references to the making and use of iron. Tubal-Cain was an iron worker (Gen. 4:22). Clearly if iron was mentioned very early in the Old Testament, then it has been used for a long time. Egyptians made wrought iron before the Exodus, David prepared large quantities of iron for the temple (1 Chr. 22:3, 29:7) and the merchants of Dan and Javan brought it to the market of Tyre (Ezek. 27:19). Many things were made of iron (Deut. 19:5, 27:5; Josh. 17: 16,18; 1 Sam. 17:7; 2 Sam. 12:31; 2 Kings 6:5,6; 1 Chr. 22:3; Isa. 10: 34). In Proverbs 27:17, we learn that "iron sharpeneth iron" indicating that iron was fundamental for industrial and military purposes. It also indicates that the ancients were aware that there were different hardnesses of iron, depending upon the source ore, smelting processes and impurities.

Iron was smelted in Egypt, Mesopotamia and China 5,500 years ago, in India 4,000 years ago, in the Caucasus 2,000 years ago, in northern Nigeria 2,000 years ago and the Hittites used iron in the 14th to 12th centuries BC. Egyptian texts dated at 3,500 BC refer to iron. The British Museum has iron tools from the pyramid of Kephron built about 3,700 BC. To build pyramids, copper or bronze tools with diorite hammers were not good work tools. The early Egyptian stoneworkers probably used harder iron tools that have since rusted out of existence. The Egyptians were prejudiced against iron for sacred and aesthetic reasons. Iron rusted and was not as beautiful as copper, bronze or gold. Iron smelting underwent technological advances in Egyptian times and about 1,500 BC in the time of the pharaoh Thothmes III (or Thutmosis; sometimes referred to in the 19th and 20th Centuries as the Napoleon of Egypt), pyramid decorations show that bellows were used for iron smelting. Bellows blast air into a hearth to give higher temperatures and better quality iron. Greek and Latin authors make many mentions of iron. In Greek mythology Ulysses plunges a steel stake into the eye of Cyclops.

A remarkable textbook

In a remarkable book written in 1556 AD, Agricola (Georgius Baumann) described the smelting of many metals including iron. Agricola tells of the complete devastation of the European forests by metal smelters and glass makers. Because coal was not used for smelting, forests were cut down and wood was burned to form charcoal. Iron ores are principally oxides, charcoal (or now coking coal) was used as a source of energy and to remove oxygen (as carbon dioxide) during smelting. This is the chemical process of reduction.

If we are to use iron today, we can either clear fell our forests to make charcoal or we can use coking coal, coal seam gas, oil, petroleum gas or hydrogen as the reducing agent. With technological improvements, there is now no need to clear fell forests for the production of iron. It is technology that has saved the forests, not the greens.

Furthermore, forest timber was burned to ash and the ash was mixed with river sand and salt (or soda) to make glass. The heat needed to melt materials for glass making came from wood. Once glass makers in the Middle Ages had completely destroyed a forest, they just moved on to the next one. The forests have now regrown and are larger and far healthier than they were 600 years ago, again because of technological advances in glass making. Glass was absolutely necessary because during the Little Ice Age (1300-1850 AD) as it was used to let in the light and retain the heat in dwellings.

What would you prefer? The use of "renewable" energy such as wood to make steel for your stainless steel teaspoon or the use of coal? What would the greens recommend? Hint: Are the forests of China big enough to make 2 million tonnes of steel a day (the current Chinese production)?

Agricola's book was translated from Latin in 1912 by Herbert Clark Hoover and his wife Lou Henry Hoover before Hoover's 1929-1933 term as US President. It was published in the London journal called *Mining Magazine*. Hoover was a mining engineer who not only worked in the US but, as a single man, worked at Broken Hill and north of Kalgoorlie in Australia. What US President today could even speak a second language let alone translate a classical text?

In Agricola's book, the early methods of iron production are described. Iron smelting was a variation on copper smelting. A furnace 1 metre high on a 2 metre x 2 metre base had a central crucible. Heat and reduction were from charcoal and the temperature was increased with bellows. A charge of iron ore, limestone and charcoal was cooked for

8 to 12 hours and 2 to 3 centumpoudia (i.e. 60 to 90 kilograms) of iron was produced.

Smelting by reduction produced malleable iron, the second process produced cast iron and a third process produced steel by cementation where crude iron was cooked with charcoal for up to a week in long stone ovens. At 1,150°C during copper smelting, the iron ore flux can form blooms of poor quality iron. These blooms were hammered into wrought iron to drive out waste materials from the copper smelting. It was these primitive hearths described by Agricola that produced bloom iron. At all iron works were ironsmiths hammering bloom iron into better quality iron.

If the iron blooms contained more than 2% carbon, then the iron could be poured into moulds to make pig iron. Vats of molten iron were stirred to change the carbon content in a process called puddling and Agricola describes how iron rods were dipped into molten iron vats, hammered on an anvil, cooled in water, heated and hammered again until better quality steel was produced. This process of steel tempering reduces the amount of impurities, rids the steel of slight imperfections in crystals, closes up cracks and hardens and strengthens the steel. This process of steel tempering is still used today.

Pre-industrial age iron

In former times, ironsmiths produced steel by trial and error. They did not understand the chemistry and metallurgy of iron production. Each province produced a different type of steel depending upon the source materials used. In a furnace, the burning of charcoal, coke, coking coal, gas or petroleum produces heat and carbon monoxide. The hot carbon monoxide strips oxygen from iron oxide to form carbon dioxide and metallic iron. The limestone flux releases carbon dioxide and the remaining lime reacts with silica to give the main constituents of the light liquid silicate waste (slag). Ancient smelting produced blooms of iron in batches whereas the modern furnace is a continuous process that produces iron of constant quality. The historical driver of iron and steel metallurgy was military technology. Great armies could be created by using steel.

Wrought iron

Iron was produced from primitive furnaces and open hearths thousands of years ago in Africa, China, UK and Europe. The fuel used was charcoal and iron production was limited by the rate which forests could be cut down and regrow. The iron was poor quality, carbon-rich, brittle and, unless it had a large amount of slag impurities, could not easily be worked.

Wrought iron was widely used before effective methods of steel making were developed. Wrought iron was produced by direct reduction of iron oxide to iron metal by charcoal in numerous small manually-operated local bloomeries. Bellows often added air to increase the furnace temperature and to speed up production of blooms. Because iron was not completely melted in the furnace, spongy blooms of iron formed. Because the iron had not been completely liquid, it did not contain large quantities of carbon from the charcoal fuel.

Wrought iron was formed from iron produced in a furnace that then underwent puddling. Liquid metal in the hearth of the furnace was not in contact with charcoal, coke or coal and the hearth was lined with iron oxides. The liquid iron was stirred with iron bars and blasted with air that oxidised carbon to carbon dioxide in the iron. Slag floated and iron particles solidified into spongy wrought iron balls. These hot balls were hammered, slag was squeezed out and the balls were rolled into sheets. Once a batch of bloom iron was produced, hot blooms were hammered into shape by a blacksmith. Some wrought iron was used to make steel. Wrought iron has a very low carbon content and contains zillions of fibrous inclusions of slag. This gives wrought iron a grain and allows it to be hammered, drawn and welded. By contrast, cast iron breaks if hammered.

Probably the best-known wrought iron structure is the Eiffel Tower, built in 1889 for the Paris Exposition. It was a temporary entrance to the Exposition grounds some 325 metres high and 7,300 tonnes of wrought iron was used for its construction. Other iron structures have been with us since antiquity. The 7 metre-high iron pillar of Delhi is at least 1,000 years old. Despite the Delhi climate, it is corrosion resistant because of the high phosphorus content of the iron. It is probable that phosphate minerals were used as a flux for iron smelting. In modern steels, the phosphorus content of the iron ore must be exceptionally low as phosphorus steels are very brittle.

The first use of coal

A thousand years ago, the Chinese started to use coal rather than charcoal for the production of iron. This simple technological change spared more forests than any environmental activism has done in modern times. Early in the 18th century in England, Abraham Darby, of Shropshire, began to fuel blast furnaces with coke. Darby, a founder and the third of that name to run the family business, had used charcoal as a fuel until he visited one of his customers, a hops producer in Kent and found that the oast houses used for drying hops were fuelled by coke. He took this technology home with him and thus history was made.

Incidentally, the period of the Industrial Revolution (mid-18th to mid-19th Centuries) coincided with the rapid expansion of Britain's maritime empire. Early in this period the production of charcoal for industrial use and the building of Britain's vast wooden mercantile and naval fleets came close to completely deforesting Great Britain. The introduction of coke and the subsequent use of iron and steel in ships saved more trees than whole phalanxes of chanting greens.

A visit to Coalbrookdale, the centre of Darby's enterprise, and the nearby Iron Bridge, the pioneering 30 metre cast iron arch bridge built half a decade before Arthur Philip sailed for Botany Bay, reveals that a place that was once described as like "entering the gates of Hell" reveals, today, a sylvan setting of rare beauty. Nature is pretty good at healing itself.

When coal is heated, it releases volatiles and a porous rigid carbon-rich residue remains. The volatiles (coal gas) were used for heating, lighting and cooking. Coal gas has now been replaced by reticulated electricity and reticulated natural gas. Coke is strong and can carry a greater weight in a furnace than charcoal but it has more impurities (especially sulphur) than charcoal. This one simple step of substituting coke for charcoal triggered the Industrial Revolution and saved the forests of England, Europe and the US. Coke has now been replaced in furnaces by coking coal. As technology evolves, the energy consumption of furnaces has been reduced. Heat, solids and gases are now recycled and fuels such as shale gas have now replaced coking coal in some furnaces in the US.

There is no escape from reality. If we are to make and use steel, then we need to burn fossil fuels. And to make steel, we release carbon dioxide into the atmosphere and embed energy in the steel.

Modern steel making

It was realised in the 19th century that if you want something big and strong, then steel is the material to use. In many parts of the world are marvellous iron and steel structures such as bridges. These are architectural memorials to the great Industrial Revolution of the 19th century and were built of wrought and cast iron. However it was not all rainbows and lollipops. During the 19th century a number of well-publicised disasters were caused by the collapse of iron railway bridges and the failure of railway tracks. This was because of a lack of quality control and the manufacture of poor quality steels. In former times, the decarburisation of pig iron was a very tedious, dangerous and inefficient process that involved stirring and agitating molten iron by hand in the presence of air. There was little quality control.

In 1855, Henry Bessemer perfected a process that allowed the fast manufacture of inexpensive, mass-produced pig iron that could be converted to wrought iron, cast iron or steel. This was a great improvement from the blooming and open-hearth furnaces of the time. The Bessemer process uses coke (derived from coking coal) and removes impurities from the iron such as silicon, aluminium, manganese and carbon by oxidation and slag formation. Air is blasted into the furnace hence the smelter is called a blast furnace. The heavier iron is tapped from the bottom of the furnace and the floating slag is tapped higher up in the furnace. The iron is poured into moulds. Most steel used in the 20th century was made by the Bessemer process. In order to operate the blast furnace efficiently, the charge needs to be constant in particle size and composition. This is why there are long-term contracts for the supply of iron ore of constant composition and size (e.g. Pilbara Blend iron ore from Western Australia).

Nowadays, the Bessemer process has been improved by blasting hot oxygenated air into molten pig iron giving greater quality control on iron chemistry. The added air, or oxygenated air, enables chemical reactions to take place throughout the furnace as a material gravitates. Scrap is often added and a lance delivers oxygen into the base of the furnace. The carbon in the pig iron is burned off as carbon dioxide and refractory bricks line the furnace. The fat belly of a blast furnace is the point at which the descending charge expands before it is melted into iron and slag. Pig iron from ladles is pre-treated to remove sulphur with a spectacular heat-emanating reaction using powdered magnesium to form magnesium sulphide. A scum of magnesium sulphide is skimmed off. Iron oxide scale is used in the same way to remove excess silicon and phosphorus.

The fluxes are most commonly limestone and silica and sometimes small amounts of serpentinite are added. The proportion of these materials is calculated very carefully to maximise efficiency, maximise slag fluidity and minimise waste. The charge to the blast furnace is continuously supplied through the top of the furnace and hot air or oxygenated air is blasted into the lower part of the furnace. Sometimes fluorspar (calcium fluoride) is added to keep slag fluid. This can have negative effects as fluorine is a wonderful flux and can induce the refractory furnace bricks to melt. Fluorine and some fluorides are highly toxic and need to be scrubbed from gases released during smelting.

More fluxes are added to remove the remaining deleterious elements, the blown liquid metal is tapped, as is the liquid silicate slag. Steel produced by this method has low carbon, manganese, silicon, sulphur and phosphorus. The iron produced from a blast furnace contains carbon and, when it is solid, is a brittle and unworkable pig iron. It is the intermediate material cast into pigs for the subsequent production of purer iron and steel. Pig iron has 3.5 to 4.5% carbon, steel has 0.07 to 0.13% and wrought iron 0.05 to 0.25%.

Furnace linings must withstand high temperatures and must not chemically react with hot materials in the furnace. Conditions in a blast furnace are extreme and furnaces are lined with refractory bricks composed of magnesium oxide, aluminium oxide or natural refractory minerals such as andalusite (aluminium silicate; Al_2SiO_5).

Magnesium oxide is made by sintering the mineral magnesite ($MgCO_3$), a process that involves a huge amount of energy and the release of carbon dioxide into the air. Aluminium oxide furnace bricks derive from the mining of bauxite and the conversion of bauxite using caustic soda into aluminium hydroxide which then is sintered into aluminium oxide. These processes also involve the use of large amounts of energy. Refractory minerals such as andalusite form naturally at high temperature so can withstand the high temperature inside a furnace.

Exploration, mining, transport and manufacture of andalusite into bricks involves a large amount of energy and emissions of carbon dioxide. Furnace bricks need to have a higher melting point than the hottest temperature in the furnace, they must not react with the furnace charge, must be physically strong and must be able to expand and contract evenly without distorting the inside of the furnace. A stainless steel teaspoon could not be made without furnace bricks.

The waste gases are scrubbed to remove fine particles and sulphurous gases. In some over-regulated green jurisdictions, some steel plants also remove some carbon dioxide from the waste gases. Heat from the waste gas and furnace is used to pre-heat the air blast, again showing contrary to green propaganda that industrial processes strive to achieve maximum energy efficiency and minimum waste. For every tonne of steel produced, 1.7 tonnes of carbon dioxide is released.

Some iron is lost to slag. Slag is cooled and this silicate-oxide mixture is used for skid-resistant road surfaces, concrete aggregate and landfill. In some parts of the world, phosphorus-rich slags have been used as a fertiliser when phosphorus minerals were used as a flux or were present in the iron ore.

There are a number of different steel making processes now. About 70% of all steel is made from iron ore using a conventional blast furnace with the modified Bessemer process. About 30% of iron is made from scrap via a blast furnace or other specialised smelting processes. These are smaller scale processes such as the direct reduction furnace, rotary kiln, rotary hearth, shaft furnace and fluidised bed. Pig iron and sponge need to be converted to steel by a batching process in a basic oxygen furnace, electric arc furnace or induction furnace. The basic oxygen furnace is used to make most steels whereas the other processes are used to make specialist alloys.

As a result of 150 years of metallurgical experimentation, research and development, we now have steels that are super hard, have extreme tensile strength, have high vibration damping, can survive Arctic weather, can be used for submarine pipelines and do not corrode. We still also have the earliest types of iron such as pig iron, wrought iron and high carbon steels.

Rusting of iron

Iron and steel rust. This is a chemical reaction between iron and the oxygen in air, especially moist air. Today rust costs us billions of dollars each year. Because iron rusted so readily, experimentation into rust-free steels took place in the 19th century. This led to the discovery of stainless steel and the galvanising of iron.

In galvanising, a thin coating of zinc metal covers iron and steel. A very thin zinc oxide coating forms on the galvanised zinc surface. This prevents oxygen and moisture from further chemical reaction with zinc. By promoting an oxidation reaction on the surface of the zinc coating, the steel is protected from rusting. Zinc is a sacrificial metal protecting steel from rusting by slowly reacting and dissolving. After many years, galvanised steel needs to be re-galvanised.

The Iron Age

The Iron Age started thousands of years ago. We are still in an iron age. So many things we use such as cars, buildings, factory products, ships, nuts, bolts and screws are made of iron. Without iron, we would be living the life of our ancestors thousands of years ago. The steel girder, has allowed us to build large and high buildings cheaply. Buildings can only be tall if steel girders or steel-reinforced concrete are used. These buildings sway up to two metres in the wind and earthquakes. Many steel girder buildings in Tokyo have survived very intense earthquakes. These buildings can be constructed easily. However, to maintain the structural integrity of the building, the iron girders cannot be weakened by rusting.

The stainless steel we use in cutlery in our everyday life derives from thousands of years of experimentation, technological advances and, more recently, a scientific understanding of process chemistry. The processes used to make iron in past times have continually been replaced by cheaper and better production methods. Each new technological advance lowered costs, lowered energy use, produced less environmental damage, reduced risk, increased safety and increased efficiency.

Can someone please tell me what industrial process that reduces energy use, reduces environmental damage and make the world a better place in which to live was invented by the greens?

5

SHINY METALS FOR YOUR SPOON

Chromium – the mantle metal

Some 70% of all chrome mined is used in alloys. Alloys, like the stainless steel used in a teaspoon, are metal mixtures. The remaining 30% of chrome produced is used in the chemicals industry for tanning, pigments, electroplating and timber preservation.

There are two states of chromium. Natural chromium in rocks, minerals, soils and waters is the reduced form (chromium III). The reduced form of chromium found in nature is neither soluble nor poisonous. Oxidation processes on Earth are normally not intense enough to produce the highly toxic chromium VI. Chromium III is oxidised by industrial oxidation processes to chromium VI.

Toxicity of chromium

If you wear shoes, have a leather handbag, have a leather jacket or stop your strides falling down with a belt, then you are an active participant in the global mining and chemicals industry because almost all leather is tanned. Old tanneries are commonly highly contaminated sites. Chewing tanned leather or treated timber and burning treated timber can create severe brain damage because the oxidised form of chromium (chromium VI) is highly soluble and poisonous.

Harvested softwood timber is impregnated with copper chromium arsenate in a high-pressure high-temperature vessel. This treated timber needs to be used sensibly. If for environmental reasons you prefer to use harvested softwood timber rather than native forest hardwoods, then you are also an active participant in the global mining and chemicals industry. Treated timber is not very tasty for insects, especially white ants, and treated timber is used for fence posts, structural support for domestic buildings, garden edges and many other domestic uses.

Burning treated timber vaporises arsenic and spreads a toxic form of chromium. There have been some shocking cases of poisoning of children from the domestic misuse of treated timber. It might burn easily and create pretty coloured flames in your cooking or heating fire but you can also poison those who inhale the smoke and fumes. It's not very smart to build garden settings and children's sand pits out of treated timber. Be one with nature and give ants and insects something to eat.

In some developing parts of the world, the chrome chemicals industry has left a dreadful mess of widespread chromium VI contamination in soil and water. Natural chromium III is synthetically oxidised to chromium VI and for each tonne of chromium VI produced, some 2.5 tonnes of chromium VI-bearing waste is produced. Rain is slightly acid, it leaches waste dumps containing minor amounts of chromium VI and runoff pollutes large areas.

For example, in Liaoning Province in China, there is chromium VI in water wells. It derives from slags from a nearby ferrochrome factory. This underground drinking water from wells reached 20 milligrams per litre of chromium VI, 200 times higher than the maximum level allowed in drinking water. Mortality rates from stomach and lung cancer are high in these areas.

Why doesn't the chromium in your stainless steel teaspoon poison you if chromium VI is so toxic? Your stainless steel teaspoon contains many elements that, in certain states and concentrations, can be toxic. The chromium in your stainless steel teaspoon is chromium metal (chromium 0) and not chromium VI. Are chromium and the other heavy metals in your spoon going to kill you?

There have been many tests on stainless steel consumer products and medical devices to determine if heavy metals are released during normal use. Studies on release of chromium and nickel from kitchenware made of stainless steel have been inconclusive. Maybe some strong food acids such as acetic or oxalic acid will release traces heavy metals from stainless steel. However, when stainless steel implements and bowls are used for food preparation using strong food acid, there is no demonstrated increase in heavy metals in the foods.

Some heavy metals may be released from stainless steel medical implants or orthodontic appliances although results are inconclusive and the released metals are well below dangerous dietary intake levels. Experiments have also checked whether heavy metals are released from stainless steel after long exposure to sweat, blood or gastric fluids or by inhalation. These experiments show that a little iron is released and minute traces of chromium and nickel can be released in very extreme environments.

Why is stainless steel stainless?

Chromium metal is the key to stainless steel. Chromium metal in nature very quickly bonds firmly with oxygen and this high affinity of chromium for oxygen allows the stainless steel alloy to form a stable, extremely thin chromium oxide film at the surface that stops corrosion. The chromium oxide film is impervious to water and air hence it protects everything covered by the film. The film is so thin that it does not affect the lustre of stainless steel. This film is called the passive oxide layer (passivation layer) and it forms instantaneously when the stainless steel is exposed to the oxygen in air.

The passivation layer not only forms in air but on medical implants and orthodontic appliances in the human body. All metals, except gold, platinum and palladium, corrode spontaneously when in contact with air. Stainless steel is also self-healing and rebuilds when the oxide layer has been removed. The nickel preserves the internal structure of steel.

A cheaper metal such as manganese can also do the same job as nickel but not as well.

However, in only the most hostile environments can stainless steel be attacked. A domestic kitchen sink is wiped down regularly with water. This is why the stainless steel remains bright because any possible corrosive agent is removed before it can attack the protective oxide layer. If a sink is wiped down with abrasives or chloride-rich detergents, the oxide layer is removed and reforms immediately. In high stress hostile environments, stainless steel is regularly cleaned to lengthen its working life.

The conclusion is that we use stainless steel because it self-repairs, it can survive all sorts of hostile conditions and not dissolve or release heavy metals that, in high doses, could be toxic. All this is because of the passivation layer. I can guarantee that if you use a stainless steel teaspoon, you will be dead within 147 years. But your death will not be due to the metals in cutlery.

Where does chromium come from?

Chromium metal is not produced directly from chrome ore (chromite; $FeCr_2O_4$). Chromite is not a typical crustal mineral because it needs high temperature for formation and is very common in the deeper hotter parts of the Earth such as the mantle. The atoms in chromite are packed very tightly because they have been pushed closer together in the high-pressure mantle. Diamond forms at more than 150 kilometres depth in the Earth and is commonly associated with chromite. Diamond also has the carbon atoms packed tightly.

Extraordinary geological processes are required to bring a deep mantle rock to the Earth's surface. One of these processes occurs when continents collide and the ocean floor rides over a continent (e.g. Papua New Guinea) rather than getting pushed underneath (e.g. The Andes, Japan). A slice of the ocean floor and the mantle beneath ends up being pushed over the land surface. In places, older rocks end up on top of younger rocks, the opposite of what we commonly see and deduce intuitively.

Another process of bringing mantle rocks to the surface is by the partial melting of a large volume of the mantle. The lighter molten rock rises to near surface and slowly solidifies into a large layered saucer-shaped mass. This melting of large volumes of mantle rock is commonly triggered by an asteroid impact with the melting in the mantle induced by a sudden loss of pressure due to shock rebound.

When you use a stainless steel teaspoon, the chromium in your teaspoon was only made accessible on the surface of the Earth by one of two unusual and extreme geological events. Because almost all of the chromium mined comes from large saucer-shaped masses of rock that were a mantle melt, you can only feed yourself with stainless steel cutlery because of an asteroid impact in the geological past.

Slices of the old ocean floor

Slices of the old sea floor and the upper part of the mantle have been well studied in Cyprus. There is a sequence that is repeatable all over the world, whether the thrusted sea floor is in the Urals, Oman, California, eastern Australia or Cyprus. The uppermost part of the sequence comprises flinty silica-rich sediments that form from the silica shells of minute floating organisms. When these organisms die, their shells sink to the ocean floor. Furthermore, these organisms can mutate very quickly with time, slight changes in ocean temperature, salinity or light. Misguided creationists look at terrestrial mammals to try to disprove evolution and ignore the rapidly-mutating floating organisms that can be used to tell when rocks formed and to show hundreds of millions of years of constant evolution. These siliceous shell oozes accumulate and are later compressed into rock.

Most floating organisms have shells made of calcium carbonate and when these shells sink, they dissolve in ocean water at about 3.8 kilometres water depth. This is an important part of the carbon cycle. The oceans contain monstrous amounts of carbon dioxide that is dissolved from the atmosphere as dissolved carbon dioxide, bicarbonate and carbonate. Slight chemical, temperature and pressure changes in ocean conditions change the proportions of them.

Back to the ocean floor. Beneath this flinty silica-rich sea floor sediment is pillow basalt. This is a volcanic rock that was erupted at mid ocean ridges. Mid ocean ridges are where the ocean crust is being pulled apart by plate tectonics, deep fractures form, the centre of the mid ocean ridge drops to form a rift valley, the depressurisation of the high-pressure high-temperature mantle induces partial melting of it and the lighter molten rock rises and erupts on the sea floor. The molten rock is at about 1,100°C. The molten rock rises because melts are lower density than solids and because they contain large volumes of dissolved gas, mainly water and carbon dioxide.

There are at least 3.47 million submarine basalt volcanoes that are not on mid ocean ridges and 64,000 kilometres length of mid ocean ridge volcanoes. There are only about 1,800 terrestrial volcanoes, most of which are non-basaltic and most exhale small amounts of carbon dioxide. Basalt volcanoes emit orders of magnitude more carbon dioxide before, during and after eruptions. These all release carbon dioxide that dissolves in the deep cool high-pressure ocean waters. Furthermore, carbon dioxide leaks out of extinct seafloor volcanoes and the fractures that criss-cross ocean floors. In places, there are pools of liquid carbon dioxide on the sea floor. A slight change in the rate of submarine volcanicity thousands of years ago would result in changes in the rate of ocean degassing today.

There can be no scientific discussion of the relative impact of human emissions of carbon dioxide without a far better understanding of natural emissions of carbon dioxide, carbon recycling, processes on the ocean floor and processes beneath the oceans. This discussion has not taken place.

Basalt melts erupted onto the sea floor roll as balls along the sea floor. The ball contains a frozen crust and a liquid centre, the hot crust reacts with seawater and piles of these balls are flattened into pillow shapes. These eruptions have been filmed many times. It is really spectacular when basalt lava at 1,100°C flows into seawater at 2°C on the ocean floor. The erupting mid ocean ridge is cooled by circulating seawater which adds some components to the basalt (e.g. sodium, potassium, sulphur, water) and extracts others (e.g. metals) and this metal-rich hot fluid is exhaled into the mid ocean ridge rift as hot springs. These metals can accumulate just beneath and on the sea floor to form ore deposits and have been observed currently forming on mid ocean ridges. On Cyprus, ancient ore deposits in these settings have been mined for thousands of years.

Underneath the pillow basalt is a sequence of more pillow basalt cut by wall-like bodies of basalt and layers of basalt that never reached the sea floor. These are the feeder zones to the pillow basalts. Beneath these basalts are masses of coarse-grained rocks of basalt composition, these were the chambers of molten rock that were erupted on to the sea floor. This sequence of rocks is the oceanic crust, it had been geologically mapped in Cyprus and many other places before ship-mounted drilling rigs drilled through the ocean crust. This drilling into modern seafloor crust validated what had been determined from previous onshore mapping of ancient ocean crust. This is why mapping is so important in geology. It is a data collection and validation process in science. The structure of the ocean floor was one of the many pieces in the jigsaw that led to the concept of plate tectonics.

Beneath the ocean floor basalt is the mantle. Well before deep drill holes were able to sample the mantle, geologists had a pretty good idea of its composition. Earthquake shock waves have different velocities depending upon the material through which they pass. These waves suggested that the mantle was composed of silicates of magnesium, iron and calcium. This is also the composition of stony meteorites that derived from the mantle of a fragmented proto planet.

Some basalts that erupt on continental areas carried inclusions or bombs of magnesium-iron-calcium silicates indicating that as the molten rock rose from the upper mantle, it plucked off some of the surrounding material. Slices of the old ocean floor on Cyprus are underlain by masses of magnesium-iron-calcium silicate rocks that were correctly interpreted as mantle rocks. These mantle rocks occur in places such as Cyprus, Turkey, Greece, Albania, Finland, Philippines, Ural Mountains, USA and Australia. They contain disseminated chromite and, in places, pods and lenses of chromite are also present.

These chromite pods and lenses contain a type of chromite called metallurgical chromite. It is this chromite that is favoured for making ferrochrome used in a stainless steel spoon because it is high in chromium and iron, low in aluminium and can be sold as lumpy ore. Fewer contaminants need to be removed as slag in smelting. Lumpy ore allows hot air to be blasted up a smelter whereas fine grains of chromite can clog up the works. The mantle masses of chromite are very difficult to find because they have no characteristic geological, geochemical or geophysical signature. Furthermore, they are difficult to mine because they are small, discontinuous and stretched out into a series of pods and lenses. The associated magnesium-iron-calcium silicates need to be removed before lumpy chromite can be smelted as they are a contaminant that must be removed in slag. Most of these chromite mines are artisan or small company mines in low labour cost countries.

Because the mantle rocks were stable at high pressure and high temperature, once they are near the Earth's surface they become very unstable. Flushes of seawater, mantle fluids and ground waters change the mantle rocks. New minerals form by hot water-rock reactions and most of the new minerals contain water and carbon dioxide. Some areas of ancient altered mantle rocks have been mined for magnesium carbonate (used in refractory bricks), chrysotile, soapstone, nickel, cobalt, gold and platinum. During these reactions, chromite is stripped of its chromium and ends up as magnetic iron oxide (magnetite) and hydrogen is leaked to the atmosphere. Images of the Earth from space show one big cloud of hydrogen leaving Earth.

Major chromium deposits

The world's largest chrome deposits occur in huge saucer-shaped intrusions that occur over thousands of square kilometres in area. Partial melting in the mantle, often triggered by a decrease in pressure during rebound after an asteroid impact, produces a very large volume of light molten rock that rises along fractures. When the liquid rock has reached its buoyancy level in the upper crust, it stops moving and starts to cool. The surrounding intruded rocks were cooked up to high temperature and, in a number of places, what were muddy rocks were cooked to andalusite-rich rocks which are mined to produce refractory bricks. The liquid rock chills to a solid at the margin. The very large volume of liquid rock is surrounded by refractory insulating solid material and takes a very long time to cool from a liquid to a solid.

As cooling proceeds, large convection eddy currents are established and the first solid minerals to precipitate out from the liquid rock are dense high temperature silicates (e.g. olivine). These fall to the bottom of the intrusion which started life a lens-shaped mass and, after loss of gases, collapsed to a saucer-shaped one. Layer after layer of these high temperature minerals form. At times, small squirts of liquid rock were injected into the hot solid rock layers, these can scour out existing layers and, by various chemical reactions, small amounts of chromite, nickel minerals and platinum minerals precipitate. When the liquid rock has cooled to the temperature at which feldspar and chromite precipitate, then a rain of solid chromite crystals falls and settle on hot feldspar-rich rocks. This happens time and time again and the end result is a solid feldspar-rich rock that contains many layers of chromite.

These layers are horizontal to about 7° slope, the upper feldspar-rich layers are scraped off and the solid chromite layers are selectively mined. Removal of more feldspar-rich rock exposes the underlying chromite layer for extraction. Feldspar is an aluminium mineral. Because the chromite layers are within a feldspar-rich rock, the chromite contains more aluminium and less chromium that metallurgical chromite. Because the chromite is rich in aluminium, it needs beneficiation and pelletisation before it can be smelted to provide the chromium to make your stainless steel teaspoon.

Cooling continues and the minerals typical of basalt rain down from the cooling liquid rock. This material forms the bulk of the saucer-shaped intrusion. The last stage of cooling is the solidification of the last dregs of the liquid as granite and the loss of large amounts of gas, mainly steam and carbon dioxide. The whole complex process of cooling has been replicated in experiments.

The best-known large saucer-shaped mass is the Bushveld Complex of South Africa. It supplies most of the world's chromium, most of the world's platinum group elements and some of the world's vanadium, nickel, titanium, iron, tin and andalusite. Russia also has a very large one.

If South Africa fell off the world, then there would be a great shortage of the basic commodities necessary for the industrial world. If Russia bought major reserves of these platinum group elements in the Bushveld Complex that are so vital to keep the wheels of industry spinning, then Russia could control the world's petrochemical, chemical and automobile industries. On top of this, if Russia could control gas supplies to Europe, then they would control industry in the Western world. Control of strategic resources is economically far more powerful than having a strong army.

Ferrochrome

Because chromite forms at a very high temperature and the iron, chromium, aluminium and oxygen atoms are very tightly bonded together, the smelting of chromite ores requires very high temperatures such as an open arc furnace which produces a temperature of 2,800°C. High temperature refractory bricks line the furnace. Such furnaces consume huge amounts of electricity to produce the very high temperature required. It is just laughable to think that electricity produced from wind, solar, wave or tidal could be used in an arc furnace.

Furthermore, if the wind stopped blowing while the furnace was operating, the interior of the furnace would freeze and the furnace would have to be rebuilt at huge capital cost. A modern society cannot be run on unreliable ideological electricity as reliable large base load sources of electricity are required to operate arc furnaces to treat the chromite necessary for the manufacture of your stainless steel teaspoon.

The Perrin Process involves melting chromite ore with lime. The lime operates as flux in the furnace and is produced by heating of limestone to release carbon dioxide. The lime combines with impurities of minor calcium aluminium silicates (feldspars) and magnesium silicates (pyroxenes) in the chromite ore to form slag. The melt is mixed in ladles up to six times with ferrochromium silicide (made from the slag) to chemically reduce the melt. Ferrochromium is tapped from the base of the ladle and slag from higher up in the ladle.

This slow, multiple-handling expensive process has been improved making it more continuous by using a molten metal bath as an anode and the roof of the furnace as a cathode. The chromite feed of 48 to 50% chromium oxide, 17% iron oxide, 15% magnesium oxide, 10% aluminium oxide and 5% silica is converted via ferrochromium silicide (37% chromium, 21% iron, 40% silicon) and then to ferrochrome containing 55% chromium, 43% iron and 1% silicon. This is a very clever energy-intensive way of producing the ferrochrome intermediate product used for the manufacture of your stainless steel teaspoon.

Most of the world's ferrochrome is produced in South Africa, Kazakhstan, India, China and Russia. Ferrochrome can only be made in countries where there is an abundance of chromite ore and cheap reliable electricity. Ferrochrome is then transported around the world to furnaces that make stainless steel. Over 80% of the ferrochrome produced is used to make stainless steel and the rest is for the chemicals industry.

Nickel – another mantle metal

The Earth's mantle is rich in nickel and the Earth's core is an iron-nickel alloy. The crust is greatly depleted in nickel. All nickel that we use comes from the mantle by a number of unusual geological processes, which is why nickel is not abundant on the surface of the Earth. In what were mantle rocks, nickel is mined as sulphides. However, if mantle rocks with no sulphides are left at the surface of the Earth to enjoy a long period of tropical weathering, then nickel accumulates as oxide/oxyhydroxide lateritic masses at the base of a thick tropical soil and can potentially be profitably mined.

Nickel sulphides

Nickel sulphide deposits occur in two major settings. Partial melting of mantle rocks creates nickel-bearing melts that rise into the crust. This melting can occur in hot spots in the mantle or by rebound of the Earth after impacting by an asteroid. The Earth has a long history of impacting, especially early in its history, and some of the world's major nickel sulphide deposits probably occurred associated with asteroid impacting (e.g. Sudbury, Canada).

The melts are squirted many times into the crust to form large saucer-shaped intrusions that start to solidify inwards. The solid edge to the intrusion allows the melt to cool internally. High temperature minerals crystallise first in this huge volume of cooling melt. These sink to the floor of the intrusion, the intrusion becomes layered as a result of minerals sinking into layers and from new squirts of molten rock. Such intrusions can contain layers rich in chromite, vanadium-bearing magnetic iron oxide and titanium oxides (e.g. Bushveld Complex). In places, these intrusions contain nickel sulphide minerals (e.g. Sudbury).

This presents a problem. Molten nickel sulphides and molten silicates don't mix and the heavier liquid nickel sulphides sink in melts. Because these melts form in the mantle tens of kilometres beneath the surface of the Earth, the heavy nickel sulphide melt should have sunk and remained in the mantle. There is evidence from mantle samples and earthquake shock waves through the Earth to suggest that there are probably sulphide layers in the mantle. Some stony meteorites contain nickel sulphides. The problem remains. If nickel sulphide and silicate melts behave like oil and water, how do we get nickel sulphides at the surface in rocks that were once molten?

The solution to this paradox is that, although mantle melts contain in the order of 0.5% nickel and upon melt solidification the nickel should end up in a silicate mineral (olivine; $[Mg,Fe,Ni]_2SiO_4]$), the sulphur is only added to the melt when it is in the upper crust. This happens with a rotten egg gas injection, the dissolving of sulphide-rich rocks that the melt may include upon ascent or even the ingestion of sulphur from dirty sulphur-rich coal and associated sediments. This explanation to the paradox has been tested. Experiments show that droplets of nickel sulphide form and sink in a melt when sulphur is added to a melt that is allowed to cook for a long time.

Some solid rocks show that nickel sulphides were as spherical droplets or occupied the interstices between bladed solid silicate minerals showing that silicate and sulphide melts did not mix. This late-stage saturation of sulphur is essential for the formation of nickel ore deposits and various geochemists and mining companies have invented chemical algorithms to show whether a mantle rock is fertile with potentially economic nickel sulphides or just contains unrecoverable nickel in silicate minerals.

The surface of ancient Earth was much hotter than now, it had a thin crust and heat from the cooling Earth was escaping. One way to transfer heat from deep down is to carry it in a molten rock that erupts on to the surface as lava. In ancient rocks 2,700 to 3,000 million years old, extremely hot magnesium-rich lavas similar to mantle rocks were spewed out, flowed in valleys and eroded the ancient land surface. This erosion took place by the lava picking up rubble and grinding away at the surface rocks and by melting surface rocks. If these surface rocks were rich in sulphides, the molten sulphides extracted nickel from the melt, these molten sulphide blobs sank and stringers of nickel sulphides accumulated at the base of lava rivers and channels (e.g. Kambalda, Western Australia).

These ancient nickel-bearing lavas were later bent double many times deep in the Earth, heated and flushed with carbon dioxide-rich hot water which changed the old original high temperature minerals into water and carbonate-bearing minerals such as talc, serpentine and magnesite. The nickel sulphide stringers were broken up, stretched and squirted into the surrounding rocks. These deposits are exceptionally difficult to find. However, all is not lost. Geological mapping, measuring the subsurface geology using magnetism and rock electrical properties and sampling from drilling give the third dimension necessary in exploration.

Exploration for nickel

In Western Australia, there had been 100 million years of tropical conditions before the current arid conditions. The intense tropical weathering leached material from the top 60 to 200 metres and these lava-hosted nickel deposits are generally covered by tens of metres of laterite (tropical soil residue), limestone, stream sediments or lake sediments. Very commonly the lake sediments are salt and salty mud. Exploration involves the geological mapping of the overlying residual material followed by rotary air blast drilling. This drilling method breaks rocks into chips and dust and blasts them up the hole where they are collected. Samples every 1 metre of residue are collected and chemically analysed by extraordinarily accurate methods (e.g. inductively coupled mass spectrometry).

A high content of nickel, chromium, cobalt and manganese in the soil, outcrop or drill chip samples normally suggest an abundance of the mineral olivine, a common mantle mineral. A coincidental nickel, copper, platinum group element and chromium soil, outcrop or drill chip anomaly suggests that sulphides may be present at depth. If the geochemical anomaly is defined by copper, arsenic and zinc, then this may indicate a basement to sulphide-rich sediments over which lava may have flowed. Massive nickel sulphide ore bodies at depth sometimes have a discrete mobile metal ion soil anomaly.

Once all this geochemical data is collected, it needs to be statistically analysed and then followed up with geophysical studies. The use of aeroplanes to measure very slight changes in the Earth's magnetism as a result of changes in near surface rocks is used to find the ancient lavas. Magnetic measurements from aeroplanes need a number of levels of mathematical processing before they can be used. Massive sulphide ore bodies that contain nickel commonly contain magnetic minerals and this magnetic mass may be surrounded by altered lava containing talc-serpentine-magnesite. This gives a huge magnetic difference between ore and the surrounding rocks. Again, this contrast in magnetism is mathematically modelled to show the boundary between ore and host rocks.

Follow up ground magnetic studies collect data using highly sensitive equipment over small areas of interest. Once an area of interest has been defined, then surface electrical geophysics is used. Again, the electrical geophysical data needs to be mathematically analysed and a model of conductors at depth is created. Sulphide ores are good electrical conductors and surface methods involve inducing the sulphide mass to create an electric current and measuring the conductivity and resistivity of rocks at depths. The combination of the geology, geochemistry and geophysics provides a target at depth that needs deep accurate drilling. The target is modelled from geophysics in order to determine the direction and depth of drilling.

Drilling can be by percussion or core methods. Diamond core drilling uses a rotating cylindrical bit with embedded diamonds. It grinds away rock by spinning and leaves a core of rock in the drill rod barrels. The drill bit is cooled and lubricated by water. When about 6 metres of core has been drilled, a clip is dropped down the core barrel, it attaches to the inner barrel which then can be pulled up with its core using high tensile steel wire. The core is laid out in trays, a new outer drill barrel is attached and the inner barrel lowered back into the hole ready to receive more core. The drill holes are internally surveyed about every 30 metres. The drill core is measured and compared to the drillers' markings to find out how much core has been actually recovered. If 6 metres of core is laid out, good core recovery would yield 5.8 metres of core. The drill core is geologically mapped.

The geological logging plots the 3D position of the depth of weathering and the different minerals, rock types and fractures. If it looks like there may be eventually a mine at the drill site, additional features such as engineering and geotechnical characteristics are recorded. Core intervals are chosen for sampling. This core is cut in half with a diamond-tipped water-lubricated saw, half the core is bagged and sent for chemical analysis. The chemical analyses are later added to a 3D model of the subsurface geology, sometimes the geophysics is also superimposed on the model. If nickel sulphides have not been intersected in the drilling of the modelled target, then down hole electrical geophysics can be used to "light up" an area around the drill hole to find conductors for subsequent drill testing by a new hole or a daughter hole from the first hole. Models and the test of the model by drilling commonly produce two very different results.

Percussion drilling has a vibrating rotating tungsten carbide-tipped hammer that breaks rocks into chips and blasts them to the surface. This is a cheap and fast method does not work well beneath the water table and can make geological interpretation very difficult. Percussion drilling can really only be undertaken when there is a good geological understanding of the area being drilled. Percussion drilling can also be used to get a hole to a depth suitable for adding a subsequent diamond drill tail. Rock chips and rock dust are collected, bagged in 1-metre intervals and stored for transport for geological logging and collection of a small representative sample for chemical analysis.

Exploration geologists use models all the time to try to understand the quirks of nature and most drill holes based on these models do not find economic ore. Very commonly the geological-geochemical-geophysical anomalies provide unforeseen surprises. Exploration is a very expensive high-risk process that involves the leading edge of many branches of science. Geologists treat models with great scepticism because exploration geology involves the constant testing and disproving of models. Exploration is a very humbling scientific process.

Is it any wonder that geologists just roll their eyes when climate catastrophists try to argue that their incomplete naïve climate models tell us that we will all fry and die sometime in the future as a result of the release of miniscule amounts of carbon dioxide? And all of this has been deduced from models and not measurement. Exploration geologists always test their models with measurement, climate catastrophists do not and cannot because they are not around long enough to be responsible for the failure of their scary predictions. One is reminded of the wisdom of Mark Twain who stated " There is something fascinating about science. One gets such wholesale returns of conjecture out of such a trifling investment of fact".

So, why do I labour with all of this? It is because some claim in the popular press that mining is just digging dirt out of the ground and are clearly ignorant of the depth and breadth of geology, chemistry, physics, engineering and mathematics used in exploration, mining and minerals processing. The public are also not aware that most exploration is unsuccessful, that it does not damage the surface and uses many remote techniques before a decision to drill might be taken. Drilling does not guarantee success. Most drilling programs are abandoned because they are unsuccessful. Green activists try to persuade the public that if a drilling rig appears on a block of land then there will be a mine. Nothing is further from the truth. However, this does not stop green deceit.

Exploration is fundamental for society. For every tonne of metal used in our modern industrialised world, we need to replace it by exploration. The mining industry creates and uses leading edge technology. And it must to remain internationally competitive otherwise someone somewhere else trumps you.

Mining nickel sulphides

If rock of interest were intersected in drilling, then more holes are planned to initially intersect the nickel sulphide mineralisation and to define the shape and metal content of the nickel sulphide mass. There needs to be many kilometres of drilling before a resource is defined and a decision to mine is made. Much of this later drilling is done from the surface and, if the nickel sulphide mass has the possibility of being economic, then a 5.5 by 5.5 metre one-in-seven exploration decline may be cut for infill underground drilling. If an economic ore body is delineated, then the decline can later be used for mining.

Cutting a decline is no simple matter and in good rock 5 to 10 metres per day can be achieved. The direction of the decline very much depends upon the rock types, fractures, weathering and layering in the rocks. The 5.5 by 5.5 metre face of rock is drilled in a pattern to allow maximum breakage. Drill holes are about 3 metres long and filled with explosives and a detonator. Different types of explosives are used depending upon how the rock is to be broken. The decline is cleared of people and machinery near the blast, explosives are set off with an electric charge and the decline is vented of explosion gases by air from ventilation bags slung from the roof of the decline.

A front-end loader lifts blasted broken rock from the face and stores it in a small offcut. The face and walls are then scaled down with a very high-pressure water jet, lose rock falls and the geologist maps the roof, sides and face. The wet rock is sprayed with fibre-bearing quick-setting concrete (fibrecrete) from a concrete truck that has reversed to the face. The fibrecrete prevents rocks falling on workers and machinery and holds back water. In unstable areas, steel meshing is bolted onto the exposed rock and the protective fibrecrete is sprayed onto the meshed rock.

Once the fibrecrete has set, a drill rig secures the fibrecrete and rock by drilling 6 metre holes, an expanding bolt is pushed into the hole, a plate is placed at the end of the bolt which is then tightened with a nut. In extremely unstable areas, a 30-metre drill hole is used for a cable bolt of high tensile steel cable which is tightened to hold hundreds of tonnes of rock in place. Nickel mines are commonly in soft talc and carbonate rocks and fibrecrete, mesh and rock bolting are generally used more than in other mines. Arches for additional support to cables are installed in the deepest parts of mines where the ground is squeezing, deforming and moving in real time. This movement of rock in underground mines is common, miners refer to this stress release as the rocks "talking to them".

After the face is safe, the drilling rig comes down to the face and drills a new set of holes to again fill with explosives calculated to blast rock to a specific fragmentation. While these holes are being drilled, a front-end loader is working behind the drilling rig and loading underground haul trucks with rock for transport up the decline to the waste rock dump at the surface. These trucks are designed to work in hot wet dark conditions and are also designed to carry anything from 30 to 70 tonnes of waste rock up a one-in-seven decline. All vehicles and people underground are in radio contact, there are designated vehicle passing bays underground and at various places underground are fireproof tubular strong refuges that contain food, water, oxygen, communications, medical equipment, medicines, bathrooms and reading material if workers underground need to sit out a catastrophe until rescue.

From the decline are cut passageways at regular intervals on each side of the nickel sulphide ore and passageways through the nickel sulphide ores. This underground development produces a small amount of ore to gain access to the main mass of ore. Large diameter vertical drill holes are drilled from one mine level to another, these are filled with explosives and slice-by-slice of the ore is blasted underground. Each blast drops anything from 10,000 to 50,000 tonnes of nickel sulphide ore.

If ore is over-blasted, waste rock dilutes ore. If the ore is under-blasted, some ore is left behind. If blasts are too big, other areas of the mine might be affected or even collapse. The natural stress and preferential direction of rock breakage is measured in mines. Blasting is a highly specialised branch of engineering. During blasting, some or all of the mine is evacuated and equipment is hidden away from the main force of the blast. Those of us who have worked underground can never forget the crippling headache derived from nitroglycerine blasting fumes when blasted areas are entered after insufficient ventilation.

If the ground is unstable and rocks are falling into the blasted space, then front-end loaders are operated remotely by radio with the operator some distance away in a protected area. Dusts from blasted ore can explode, blasted piles of ore underground can heat up, fume off acid and re-cement. Blasted ore is carried to a haul truck with a front-end loader, the truck crawls its way to the surface and dumps the ore on the run of mine ore pad (ROM pad). Waste rock is dumped elsewhere and low-grade ore (i.e. low metal content) is sometimes stockpiled or blended with high-grade ore on the ROM pad.

Mining creates spaces underground. Some ore may be left as vertical or horizontal pillars that operate as a buttress. The mined space is filled with a paste mixture of tailings from the processing plant, sand, slag and cement. Paste mixed at the surface is gravitated down large diameter pipes in specially drilled holes into open underground spaces and sets as hard as concrete after about 120 days. It is not just simply a matter of dropping a slurry from the surface to underground. During the fall of the paste from the surface, components can separate and there is a whole area of study that deals with turbulence of slurries and separation of components. Towards the end of the mine life, these vertical and horizontal pillars of ore can be extracted safely with the hardened paste fill operating as firm walls, floor and roof. Some pastes slightly expand after consolidation. Some spaces in areas where all the ore has been extracted are filled with unconsolidated waste rock to buttress wall rocks and, in some mines, spaces are left open. Over time these spaces will close due to natural forces underground or self mine upwards to be eventually filled by expanded rock.

Underground mines are rabbit warrens in which one can easily get lost. All mines have a second egress in case of a disaster and a number of updraft (stale air) and downdraft shafts (fresh air) for ventilation. There are storage areas for explosives (magazines), eating areas (crib rooms), toilets, gas monitoring equipment, ventilation fans, cold water dams for cooling ventilated air, waste dams for pumping water to the surface, electrical substations, light and heavy vehicle workshops, diesel fuel storage and filling areas, equipment storage bays and strict protocols for driving, access and procedures. In big mines, ore is crushed underground and hauled to the surface in skips in a shaft, by conveyor or in trucks up a decline. Underground mines operate 24/7 as a subterranean colony providing you with the basic necessities for life.

Real men and women work underground. No chardonnay socialists or café latte set in fashion clothes have ever been seen down there with Pluto in the heat, darkness, dangers and grime. Only people with a firm handshake who look you in the eye and have a great sense of humour work underground. You can count on a careless sawmiller's hand the number of greens who work underground. Maybe the work is too hard for greens? Maybe the work is productive? Maybe greens don't have practical skills? Maybe greens don't work underground because their iPhones and iPads have no reception? Maybe greens don't want to provide anything for their fellow man?

The great thing about an underground mine is that it is a green-free zone.

Processing nickel ore

The run of mine ore is picked up with a front-end loader, dropped through a jaw crusher and reduced in size. It is again crushed using a gyratory, hammer or jaw crusher to a smaller fragment size. The oversize is crushed for a third time and the undersize is conveyed for grinding in ball mills, rod mills or autogenous grinding mills.

Ball mills are large rotating steel cylinders to which cannon balls of steel are added to crushed ore and water. In some mines these balls are made of chrome steel. Rod mills use steel rods rather than balls and, if the ore is hard, it can grind itself in a semi autogenous grinding mill. Some 2% of the world's electricity is used to crush and grind rocks and, on mine sites, the greatest consumption of energy is by the grinding mills.

The slurry from the grinding mills is screened, oversize is returned for regrinding and the undersize can have materials separated in a spiral driven by gravity and water. The crushed and ground ore slurry is added to froth flotation cells. Cells are stirred, air is bubbled through the cell and chemicals added to the cell force bubbles to stick to heavy sulphides. The bubbles rise carrying the sulphides, the froth at the surface of the cell is skimmed off and the residue is sent to a tailings storage facility. Froth flotation involves multiple cells that scavenge and clean the sulphide concentrate. The sulphide concentrate contains nickel sulphides, nickel-bearing iron sulphides and copper sulphides. By passing over a magnetic drum, magnetic non-nickel containing minerals are removed, the sulphide concentrate is then agitated and water is decanted and the wet concentrate is filter pressed to remove most of the water.

Smelting nickel concentrate

This concentrate containing about 12% nickel and 1% copper is then transported to a smelter for metal production. The concentrate is roasted, sulphide minerals oxidise and the released sulphur gases are collected and converted into sulphuric acid. The metal oxides are heated with gas or coal to 1,200°C with fluxes, molten silicate slag floats on top of a liquid metal mixture of iron, nickel and copper (matte). The liquid matte is mixed with more flux, air is blown through it, iron is oxidised and the matte is left with nickel, copper, cobalt, small amounts of precious metals (e.g. gold, platinum group metals) and sulphur. The molten matte is cast into moulds and cooling is tightly controlled. The matte is crushed, ground and metals are removed by magnetic separation.

Metals undergo refining by roasting, chemical reduction and collection of nickel as a vapour with copper and precious metals remaining in the residue for further treatment. Another more conventional method is to dissolve the matte in acid and pass a large current through the solution to precipitate the nickel metal on an electrode. Another method (Mond Process) is to react roasted concentrate with hydrogen to produce impure nickel metal and steam. The impure nickel metal is then reacted with carbon monoxide to produce the gas nickel carbonyl ($Ni[CO]_4$). In the presence of excess carbon monoxide and heat, nickel carbonyl decomposes to purer nickel metal powder or metal plate and carbon monoxide.

The same minerals that cause problems with stacked nickel ore underground can be used to assist with extraction of nickel. If a dry very finely powdered concentrate is blown into a smelter, the minerals oxidise, give out heat, melt the concentrate and produce a slag that floats on a nickel-rich matte. This process of flash smelting produces heat that is used in other parts of the processing operation.

Some concentrates are oxidised then leached under pressure with ammonia to put nickel in solution. The nickel, copper and other metals can later be precipitated as pure metals. Some nickel sulphide ores are very reactive to bacteria which put nickel and iron in solution. This solution can be used to create ferronickel which is used directly in the manufacture of stainless steel for your spoon.

Lateritic nickel ores

The second type of nickel deposits are lateritic nickel deposits. They result from the long-term tropical weathering of mantle rocks that contain about 0.5% nickel. Many components are leached from the rock over tens of millions of years and the residue is enriched in iron, nickel and cobalt minerals. Such deposits are mined in Cuba, New Caledonia, Indonesia, Philippines and Australia. The fact that many of these deposits now occur in temperate latitudes is good evidence that climate has changed over the last 100 million years and that continents move.

Lateritic nickel deposits contain an iron-rich cap underlain by an iron oxide/iron hydroxide layer that can contain some nickel and cobalt. Underneath this layer is a greasy lighter-coloured rock (saprolite) that contains magnesium-rich minerals and iron oxides. At the base of this saprolite layer, nickel is commonly concentrated as oxides, oxyhydroxides and silicates.

Exploration for lateritic nickel deposits involves the identification of a mantle rock from outcrop or aerial magnetics. The soil profile is examined from water well drilling, pits, dissected landscapes and shallow drill holes. Detailed exploration is by auger drilling, percussion drilling, trenching and pitting. Samples are chemically analysed and larger samples are taken from trenches and pits for metallurgical testing.

Because lateritic nickel deposit are a surface blanket, mining first removes the upper barren layers for storage for later minesite revegetation. The soft nickeliferous layer is scraped from the bottom of the ancient soil profile, transported to the ROM pad and stockpiled for processing. As mining proceeds, rehabilitation of mined areas follows.

Processing of lateritic nickel ores is fraught with difficulty because of slight changes in the chemistry and minerals in the laterite. There have been some spectacular failures of lateritic nickel mines by both big and small mining companies. A slight change in water quality can result in precipitation of various materials that clog up the works.

Extensive testing, mathematical validation and variability studies are used to minimise problems with processing rocks. Although lateritic nickel deposits form the bulk of known land-based nickel reserves, they supply less than 50% of the nickel for your stainless steel teaspoon.

Very high-grade lateritic ores are dried and sintered using high temperature. The sinter is mixed with limestone and coke in an electric or gas-fired furnace where nickel oxides are reduced to liquid nickel pig iron. This produces a ferronickel or nickel sulphide matte which is refined the same way as mattes derived from nickel sulphide ores. This nickel pig iron can be directly mixed with ferrochrome to produce stainless steel. The high-energy costs and poor cobalt recoveries are not desirable and this process, like other metal making processes, emits large quantities of carbon dioxide into the atmosphere. Many lateritic nickel ores are now leached in an autoclave at high temperature and high pressure with acid, caustic soda or ammonia. Leaching using sulphuric acid in inert titanium, ceramic or stainless steel autoclaves extracts more than 90% of the contained nickel and cobalt.

Sulphuric acid is produced as a by-product of the smelting of sulphide ores. Problems exist with the varying solubility of iron and aluminium oxides and hydroxides and calcium sulphate, with the precipitation of these chemicals in pipes and chambers and with slow rates of purification, filtration and settling. Acid consumption can vary greatly and the metals are extracted from solution as sulphides or by electrowinning to produce nickel and cobalt cathodes. Sulphides produced by processing of lateritic nickel ores are chemically treated to produce metal powder that is later briquetted.

The chemistry, mineralogy and process engineering involved in the production of nickel for your spoon from a lateritic deposit is highly sophisticated. There is a great future for improving the processing of nickel laterites by resin-in-pulp methods, solvent extraction, heap leaching at atmospheric pressure, selective use of precipitating agents, optimising known methods, reducing the large volumes of waste materials from the back end of the processing plant and by producing chemically stable residues.

Emissions of sulphur oxides from smelting, hydrogen sulphide and ammonia emissions associated with leaching, release of nickel carbonyl, windborne dust and liquid effluents all present potential environmental problems. These all are managed and, as always, there is scope for creativeness, new technology and improvement. Don't hold your breath that greens will solve any of these industrial problems.

Abundance and production of nickel

Nickel is mined from nickel sulphide deposits (45% production) and lateritic nickel deposits (55%). There is now an increasing proportion of lateritic nickel production. Global output of nickel is divided between Indonesia (18%), Russia (15%), Canada (10%), Australia (10%), New Caledonia (7%), China (6%) and the rest of the world (34%). At current mining rates, the largest reserves of nickel have been measured in Australia (173 years), Brazil (131 years), Cuba (74 years), Indonesia (58 years), New Caledonia (51 years) and Russia (23 years) with Australia containing by far the greatest tonnage of reserves. The largest producers of nickel are Nor'ilsk (Russia), Soroaka (Indonesia), Jinchuan (China) and Sudbury (Canada). Five large companies account for 40% of the total global production of nickel.

Not only is nickel used in your stainless steel teaspoon, its corrosion resistant properties, strength and heat resistance are such that it has a diversity of industrial uses such as in alloys, plating and foundries. We all know of its uses in coins, batteries and fuel cells. Some 66% of all nickel is used in stainless steel and the demand for and price of nickel has increased because of an increasing use of stainless steel for food and liquid transport, food preparation and the building of corrosion-resistant boats, trains and machines in the developing world.

Nickel toxicity

Nickel is by far the most common metal allergen and is more often a problem in women than in men. This is probably related to the wearing of jewellery, especially earrings in pierced ears. Nickel allergies affecting pierced ears are commonly marked by itchy red skin. In 2008, the American Dermatitis Society voted nickel as the allergen of the year. As more men pierce their ears, an increase in nickel allergies has been seen. Ear piercing in not the only place where nickel is in contact with the skin. I'm not too concerned about the medical effects from piercing of the tongue, eyebrows, lips, cheeks, umbilicus, nipples and God knows where else as it is a good case of Darwinism in action.

The EU has now regulated the amount of nickel that is allowed in products that come into contact with skin yet the one and two Euro nickel-copper alloy coins give out well in excess of the EU regulations for nickel contact to the hands. Sensitivity to nickel may be present in patients sensitive to pompholyx, a common type of eczema affecting the hands. Many earrings are now nickel free.

In the US, the minimum risk level for nickel and its compounds is low (0.2 micrograms per cubic metre) for inhalation during 15 to 364 days. In nickel smelting operations, very strict safety guidelines are present because nickel sulphide fume and dust may be carcinogenic and nickel carbonyl, an intermediate product in smelting, is extremely toxic. It has a double effect due to the toxicity of nickel and the release of carbon monoxide. Just to cap it off, nickel carbonyl is explosive in air. On the bright side, nickel plays an important role in the biology of microorganisms and plants, mainly in enzymes.

Manganese

Some cheaper stainless steels use manganese rather than the more expensive nickel. Manganese is one of the top ten most abundant elements in the crust of the Earth. However, concentrations of manganese are rare. Modern submarine hot springs associated with the pulling apart of the ocean floor are generally shrouded by an envelope of manganese precipitates and the ocean floor is littered with manganese nodules that have extracted manganese, nickel, cobalt, lithium and light rare earth elements from ocean waters. Cooking up of old hot spring precipitates gives manganese silicate rocks.

The main manganese deposits mined are oxide/oxyhydroxide deposits formed from tropical weathering of carbonate rocks. The most common carbonate minerals contain calcium, magnesium, iron and manganese. If limestones with a trace of iron are deeply weathered, they give the terra rossa soils so loved by red wine drinkers. If limestone has traces of manganese and have had a long period of tropical weathering, only manganese oxides/oxyhydroxides are left at the surface. After open pit mining, the black manganese ore is beneficiated before being used as an additive in a smelter to make ferromanganese. Very little manganese is made as metal, most is added as a charge to a smelter to make alloys.

Moly

Small amounts of molybdenum may be used in stainless steel. Molybdenum is not a word you try to pronounce when you have a mouthful of dry biscuits. There is only 1 to 2 parts per million molybdenum in the Earth's crust and extraordinary crustal geological processes are required to concentrate molybdenum to economic grades of 0.1 to 0.25% molybdenum. Most molybdenum is produced as a by-product of copper mining in Chile, USA, Canada and Poland. The world's largest molybdenum producers are USA, China and Chile. These countries produce 80% of the world's molybdenum and have 80% of the world's molybdenum reserves.

The most common molybdenum mineral is the sulphide molybdenite. It is a remarkable mineral. It is a soft silvery platy mineral that bends, soils hands and is so soft it can be used for writing. Molybdenum ore is crushed with primary, secondary and tertiary crushers until ore fragments are less than 9 mm in size. Crushed ore and water are added to ball mills, ore is ground to a paste, oversize is screened off and the slurry is added to aerated agitated flotation tanks where the heavy molybdenite is persuaded by chemicals to cling to air bubbles. The surface froth is skimmed off. If the crushing and grinding of molybdenite is not efficient, then molybdenite smears on metal in the processing plant.

Dried solid molybdenite is used as a solid lubricant because the plates slide over each other and bond onto metal surfaces. The molybdenite concentrate obtained from the froth flotation is acid leached to remove traces of impurities such as copper and then roasted at 650°C to release sulphur gases (for the manufacture of sulphuric acid) and molybdic oxide containing at least 57% molybdenum. Molybdic oxide produced from roasting needs to be reduced to produce molybdenum metal. You are now familiar with this process of reduction that uses huge amounts of heat and a chemical reductant such as coal, gas, petroleum, carbon monoxide or hydrogen.

Some molybdenite concentrates contain traces of the metal rhenium. During roasting, this is released as a vapour, the rhenium oxide is leached with ammonia to make ammonium perrhenate, an intermediate material for the production of rhenium metal. Without rhenium, we would have no modern efficient turbines in jet aeroplanes and would not be able to generate electricity from gas as efficiently. The more rhenium in a turbine blade, the faster it can spin and it is a strategic metal for fighter planes.

Older long haul aeroplanes (e.g. Boeing 747) required four engines but with rhenium-rich turbines a two-engine long haul Boeing 777 can cover the same distances far more efficiently with bigger and faster turbines. Each year only about 75 tonnes of rhenium is produced. Rhenium is almost impossible to break, it is heavier than gold and red-hot glowing rhenium turbine blades at nearly 3,000°C are still strong.

Molybdenum is used in the production of alloy steels and stainless steel. It provides strength and corrosion resistant properties. Some 34% of molybdenum is used in construction engineering steel, 26% for stainless steel, 13% for chemicals, 10% for specialist steels, 7% for cast iron, 5% for superalloys and 5% for molybdenum metal. Although molybdenum can be substituted for many of its uses by vanadium, chromium, columbium, boron, tungsten, tantalum and cadmium, substitutes are rarely used because of the availability and price of molybdenum. Different types of stainless steel have different quantities of molybdenum up to a maximum of 7%. Other uses of molybdenum are as a catalyst for converting crude oil to the various fractions of petroleum products and in paint pigments, flame retardants and light bulb filaments.

Stannum

Only some of the most modern stainless steels contain minor quantities of tin. The price of tin is about $22,000 per tonne and, like nickel which is high priced, substitute metals are always being tried in new alloys. Tin is a crustal metal and is not abundant. In nature tin occurs as the oxide (cassiterite; SnO_2) and less commonly as the sulphide (stannite; Cu_2FeSnS_4). There are other even rarer tin minerals. Tin is used in the alloy (bronze, copper-tin; solder, tin-lead-antimony; stainless steel) and chemicals industries, for tin plate and for anti-foulants on the hulls of boats.

EU directives will lead to the phasing out of lead from solder, the tin content has increased as a result and the growing world electronics industry needs more solder. The biggest tin producers in the world are Peru and China with minor production from Indonesia, Malaysia, Australia and Central Africa. Smaller additional deposits will come on stream in Morocco and Australia. Although tin is a strategic metal, neither the EU nor the US has an indigenous supply of tin.

Remelted crustal rocks, the last liquid dregs of a granite melt and some granites are enriched in tin. There is an intimate association between tin and granite. Granite is a feldspar-quartz-mica rock that was once a melt. It crystallised deep below the surface. The geology of tin is so well known that there is a class of granites known as tin granites. These have an arrangement of the grains showing that the melt was high in fluorine, the melt rose to near the surface (and in places may have even vented as volcanoes), that a frozen skin grew around the melt when it almost reached the surface and the melt solidified inwards. Compared with other granites, these tin granites are rich in silicon, potassium, aluminium, uranium, thorium, rare earth elements, tin and fluorine and depleted in iron, magnesium, manganese and calcium. They can be found from aerial surveys because they are low density, radioactive and non-magnetic.

During the solidification of the tin granite, large atoms such as tin and tungsten are left behind and concentrate in the last dregs of the melt. Every time there is a release of hot fluorine-rich liquids and gases, tin is scavenged from the melt into them. If these fluids are released explosively, a massive conical crater of tin-bearing broken rock forms. If the fluids are released less explosively, they hydrofracture rocks and parallel sheets of quartz containing cassiterite fill the fractures. In places, these fluids can chemically react with limey rocks, release carbon dioxide and precipitate tin minerals.

In all cases, these processes take place close to the surface and uplift, weathering and erosion exposes these tin deposits that then shed cassiterite into soils, streams and rivers. Tin has been mined from alluvial, soil and continental shelf sands and gravels and from hard rock tin mines. However, because tin occurs near the surface, is associated with source granites that are easy to find and commonly have associated alluvial deposits, most easily discovered tin deposits have been found. Very few tin deposits have been discovered in the last four decades and the price of tin has risen because of its increased uses in solder and the tightening supply due to the lack of exploration success.

Hard rock tin mining and extraction

Perhaps the tin for your stainless steel teaspoon came from China. In southern China deep underground mines exploit what were limey rocks that had reacted with hot metal-bearing fluids released from granite. At Dachang, the mine is entered via a vertical shaft on a mountainside. At the base of the shaft, workers go deeper down a steep decline on a trolley car and then go even deeper on a second trolley car. The mine is very hot, wet and poorly ventilated, workers operate with the minimum of clothing and safety equipment and the tin ore is drilled for blasting using a hand-held compressed air-driven drill. The position of the ore has been determined by underground diamond core drilling.

The ore is then blasted with dynamite (rather than modern safe explosives such as ANFO), broken ore is collected by hand and loaded into underground railway wagons by shovel. Mined ore is hauled to the surface using the trolley car and shaft system, loaded onto trucks and transported by truck down hill to the treatment building located in the village of Dachang. This tin deposit is the richest in the world, it contains 4% lead, 4% antimony, 1.8% tin and minor arsenic, zinc and silver. It has at least another 50 years of ore left at the current mining rate.

After crushing and grinding the ore in water into a slurry, the heavy lead and antimony sulphides are selectively separated by chemically persuading these minerals to attach to bubbles, the mineral-bearing froth is scraped off, dried and sent to smelters for recovery of lead and antinomy. The slurry is then passed over scores of Wilfey tables. These tables are set at a slope, have riffles and shake the slurry and water backwards and forwards. The black cassiterite concentrates on one part of the table, the lighter material elsewhere and both are selectively washed off for dumping (tailings) or drying (cassiterite). The cassiterite is sold to smelters for the production of tin metal. The electric arc smelters, mining and processing use huge amounts of electricity derived from burning coal. As with the other metals for your spoon, tin could not be produced from wind or solar power.

The dumped tailings flow down a gutter in Dachang and further downstream other factories harvest the tailings slurry and retreat it. An operation like this would not be possible in the Western world because of the underground and operating plant safety conditions and the environmental damage from spreading the lead, antimony, arsenic and sulphides all over the village and into the river systems.

The Dachang tin mines are a window into the "good old days" and what Western world mining was like in the late 19th and early 20th Centuries. Photographs taken 100 years ago at Western world mines could have been taken at Dachang today. In Western mines a century ago, comparatively little attention was paid to safety, people died like flies and huge areas were polluted. Once the Western world became wealthier, labour organised and technology applied, the mines became safer, cleaner and more efficient.

The same will happen in the developing world when they are wealthier. Environmentalism and green politics are the preserve of the wealthy in democratic societies and do nothing to help the developing world. In the developing world there are necessities such as working for food and shelter whereas in the wealthy Western countries, people can chose not to work, to live off welfare and choose their diet. In the Western world, one can even live off the work of others and, by green activism, prevent new jobs being created.

Alluvial tin mining and extraction

Maybe the tin in your stainless steel teaspoon came from offshore mining in SE Asia. There were meandering river systems with deltas that drained from tin deposits in and around the granites of Malaysia and Indonesia. Because cassiterite is very heavy, each flood event moved cassiterite downstream, lighter material was washed off further downstream and the heavy cassiterite concentrated in gutters and deep holes in the river and on flood banks.

The old river system was covered when sea level rose 130 metres during interglacials and was exposed during glacials. The river sediment was reworked and there is layer upon layer of sediment deposited in glacials (tin-bearing terrestrial sediments) and interglacials (shallow marine sediments). To find these deposits covered by up to 130 metres of seawater, systematic drilling and shallow seismic studies are undertaken to produce a 3D picture of the ancient river system.

This is essentially a 3D picture of the effects of climate change in a tropical area producing multiple events of terrestrial sedimentation, sediment reworking and cassiterite deposition during glacials and multiple events of shallow marine sedimentation during interglacials. It is this practical understanding of climate change that allows cassiterite to be discovered and mined in shallow marine offshore environments.

Once the old river system has been mapped, dredges suck up sediment from favourable sites such as the deepest parts of old rivers and flood plains. This sediment is cut with a rotating abrasive bucket wheel and sucked to the surface using pumps. The water-saturated sediment gravitates down spirals with the lighter material flung to the edge of the spiral with the heavier material concentrating in the centre. The lighter material is dumped back into the areas mined and the heavier material is collected.

The heavy minerals comprise magnetite, ilmenite (iron titanite; $FeTiO_3$), rutle (titanium oxide; TiO_2), zircon (zirconium silicate; $ZrSiO_4$), monazite (rare earth element thorium phosphate; $[Ce,La,Th]PO_4$), gold and cassiterite. The whole shipboard operation is run using diesel fuel and hence carbon dioxide is emitted.

After the heavy mineral concentrate is dried, the magnetite is removed by magnets. The ilmenite, rutile and zircon are selectively separated by using electromagnets and rotating electrostatic drums and the monazite, gold and cassiterite are separated by selective gravity methods such as shaking tables, spirals and jigs. The titanium minerals are used as whiteners in paints, the zircon is used for ceramic glazes, the monazite needs further treatment to extract the rare earth elements and thorium for the electronics industry and the cassiterite is sold for reduction to tin metal in an arc furnace.

Ancient bronze

To reduce the tin oxide cassiterite into tin metals requires very high temperatures. Today, this process takes place in an electric arc furnace and there is some discussion whether such high temperatures could have been achieved in antiquity to make bronze. Agricola describes the gravity separation of cassiterite from gravels, soils and crushed quartz from veins in Central Europe and describes the smelting of tin using powdered charcoal and bellows. However, it is not that easy to make the copper-tin alloy bronze. Hoover's footnotes in Agricola suggests that remelting the separately smelted copper and tin was the way in which bronze for the Bronze Age was made. However, the geology of the ancient world tells a different story.

The copper supply for the pre-Roman world appears to have been largely derived from Sinai and later from Cyprus. The Latin name cuprum derives from *aes cyprium* (Cyprian copper) which suggests the importance of Cyprus as a source of copper in the ancient world. The copper sulphide, copper oxide and native copper deposits on Cyprus have been mined since the time of King Solomon.

Copper has also been mined for millennia from Egypt, Syria, Israel and the Pontid Mountains of Turkey. Well, what about the tin for ancient bronze? Very minor occurrences of tin are present in Egypt and Anatolia. However, bronzes of Asia Minor and the Mediterranean contain 2 to 10% tin and there was just not enough ancient tin mining to provide the volume of tin for ancient bronzes. It might be that smelting of a copper-tin ore or stannite, the copper iron tin sulphide, may have provided the bronze for the ancients. One such copper-tin mine near the town of Çelaller (southern Cappadocia, Turkey) contains early Bronze Age artefacts.

In the Orient, copper-tin ores are common in parts of China, Thailand and Malaysia and the almost simultaneous discovery around the world of bronze probably resulted from the direct smelting of copper-tin ores. Copper-tin-arsenic ores on the coast of Cornwall (England) may have provided the raw materials for early bronze and there is one view that Homer's *Iliad* was a coded message with directions to the coastal copper-tin fields of southwestern England. In the *Iliad*, advanced smelting and working of metals is also described. Copper-tin ores in Europe (Erzgebirge, Saxony; Panasquira, Portugal) were discovered well after the start of the Bronze Age.

Your humble stainless steel teaspoon

The aim of Chapter 5 was to summarise and show just how much history, experimentation, knowledge, energy and effort goes into making the alloy stainless steel for your teaspoon. The huge and very complicated processes are just to provide the iron, chromium, nickel and possibly manganese, molybdenum and tin for your 18:8 stainless steel teaspoon. For any other item you use in every day life, a similar complicated story can be written. There is an even more complicated story to describe the processes that give you a mobile phone.

Furthermore, to explore and find the coal, iron, nickel, manganese and tin for your humble stainless steel teaspoon, geologists need a profound and practical understanding of past climate changes. By understanding the past, geologists have a better idea of the future than computer modellers. By contrast, the greens' carping about their modelled climate catastrophes provide nothing of use for their fellow man.

We take our humble stainless steel teaspoon for granted. Very few of us know the basic industrial processes to make something simple and useful and hence the greens easily slip into demonising the very materials such as coal and metals that allow us to live our comfortable lives in the Western world. Furthermore, a massive amount of energy is used, this is unseen and is embedded in your stainless steel teaspoon.

This process of exploration, mining, processing, smelting and refining did not happen by accident and is the result of thousands of years of experimentation and knowledge. We only have a patchy record of early mining and smelting and all the indications are that thousands of years ago many metals could be produced. Since then, the metal-producing processes have become more complex, more efficient, safer and predictable.

If you happen to use a stainless steel teaspoon rather than your fingers, then you are dependent upon a long chain of processes in many different parts of the world. The first process is started by geologists understanding and using climate change to find metals. To make the metals to allow you to eat, then mining, coal burning, electricity, carbon dioxide emissions, transport and international trade are necessary.

I invite the greens to invent a better process. But a warning first: this will require risk, capitalism, international trade, finance, creativity and knowledge. Don't hold your breath.

6

GREENS IN PERSPECTIVE

Yawn

A recent Gallup Poll in the USA showed that the American public just does not share a sense of urgency or a perception of the need for action on human-induced climate change. On a list of 15 items, the environment was at 14 on the list. In terms of perceived environmental hazards, Americans regard pollution of drinking water as the most important problem and human-induced global warming as the least worrisome. This confirms what surveys elsewhere show.

The average punter is interested in other issues, is tired of being lectured to from upon high and is not really interested in niche social fads that the left earnestly huff and puff about such as "marriage equality" or human-induced global warming. These are existential matters. Far more pressing and obvious issues are not even mentioned. In fact, the more preaching, marching and violence by green activists, the more the average punter rolls his or her eyes and suggests that these activists get a life. The greens think the punter is a fool, that the punter can't be trusted to vote correctly and that the greens are superior in intellect, knowledge and morality. Nothing could be further from the truth.

A Galaxy Research survey in Australia showed that an increasing number of Australians are unwilling to pay anything to fight future global warming. Only 4% of Australians are willing to pay over $1,000 a year to fight future global warming. Even when people see a need for emissions reductions, they are unwilling to pay the costs.

The punter is not a fool. The greens and their handmaidens on the left have burned off the public who have a genuine interest in the environment by their use of the social media echo chambers with vulgarity, threats of violence, sexist and racist comments, threats to freedoms, closing down of arguments, personal attacks on people the public admire and trying to win arguments by bullying and not logic.

To express a simple idea in 140 characters on Twitter is not possible. This is why it has become the medium of hatred, abuse and vulgarity. Very often such comments come from state-funded media networks, universities, green groups and anonymous unhappy souls. The greens' lust for instant self-gratification was not well thought through and they certainly won many short sniping battles but will lose the war because the public is just fed up.

The punter has noticed that the green "scientific" activists (especially those with mawkish titles) preach from their official or semi-official pulpits, gather their robes and then leave. It has been noticed that they do not engage in public debates with eminent scientists, they respond to simple questions with *ad hominem* attacks and will not answer questions despite having many spokespersons and a supportive culture in the media, public service and the green community.

The punter has not been impressed by the greens' profanity substituting for reasoned argument and has rightfully concluded that the green warmists have no arguments to support their predictions of a forthcoming global warming catastrophe. Those, like me, who disagree with the green scare campaign that human emissions will create catastrophic global warming are from widely different backgrounds, expertise, interests and perspectives and have no spokesperson. We are not organised. We just call it as it is and have been helped by nature doing what nature does.

As the global warming scare continues to collapse, some of the greens and others in the alarmist industry are busy looking for reasons for their failure to convince the public of the validity of their message. The greens just simply don't comprehend that after decades of failed predictions of a looming environmental catastrophe, people are sick of the greens crying wolf about carbon dioxide. They have a life to live.

Like all past scare campaigns driven by totalitarianism, the suppression of dissent and discussion was of paramount importance to keep the public tuned in to the green message. Recent democratic elections show that the public has tired of the greens' totalitarianism and catastrophist messages. The public is not very sympathetic to those who hate the system that actually keeps them alive.

The greens will not go away. They dominate universities, research organisations, schools, bureaucracies and the media and many live off creating scare campaigns. Many would argue that these folk have never had a real job. It will take a generation for these people to disappear from their unelected unaccountable public funded positions. However, politicians are driven by public opinion, public opinion has changed and, when it comes to political policy, the green activists are increasingly being left out in the cold.

Those popular political parties on the centre left that have allied themselves with green philosophies now stand for nothing and need to rebuild from the bootstraps up. These centre left parties used to claim that they were the parties of the disadvantaged and poor yet their actions show that they have been responsible for hurting the impoverished with fuel poverty and unemployment.

Maybe the green activists and the lunar left know the punter's view which is why they are becoming louder, more hysterical, making even more absurd claims, more vulgar and more vitriolic. Green activists seem to think that gullibility is a virtue. The activists' promotion of the leaked parts of the UN's IPPC report (AR5-WG11) demonstrates this.

The IPCC claims are now so absurd that only media organisations captured by greens give airtime to the exaggerations underpinned by no new evidence. However, denial is one of the characteristics of the green and lunar left circles and I suspect that they have become noisier because political leaders have smelt the winds of change. I call them lunar left circles because circles goes round and round, get nowhere and never progress.

A number of groups such as the Global Warming Policy Foundation and the Copenhagen Climate Consensus have shown that, with known technology, the economic cost of tackling the speculated climate change would be three times higher than taking no action. This is also shown in the latest IPCC report but, because the IPCC is now so discredited, this report is being ignored.

Former US Vice-President Al Gore has devoted much of his public life to what he calls "the greatest challenge humanity has ever faced". He is the first person to make a billion dollars from climate scare campaigns. The UN Secretary General has called climate change "the major, overriding environmental issue of our time and the single greatest challenge facing environmental regulators. It is a growing crisis with economic, health and safety, food production, security, and other dimensions". Both have discredited themselves because, over the last decades, the exact inverse of that predicted in these scare campaigns has happened. It is wealth that can solve environmental and human problems, not the green destruction of wealth.

The Gallup Poll shows one thing: the average American is not stupid. They do not want to regress into a new Dark Age so revered by the greens and they do not believe journalists who are not sceptical and close down debate. Maybe they are not prepared to give control of their destiny to the UN or to big government programs. They are aware of weather anomalies and just yawn when green activists are rolled out to tell them that this is climate change, that this is their fault and that the end is nigh.

The average American probably thinks that there are more pressing problems such as the economy, employment, education, health and dysfunctional government. Maybe the average American is just a little tired of continual dire warnings of forthcoming catastrophes such as the population bomb or Y2K that just don't seem to eventuate.

Oxymoronic green morality

Up until now, the greens have had an easy run because there was not enough time to see the consequences of their policies. We have now had a couple of decades to look at the green activists, their actions and their solutions to perceived problems. And it is not a good look. This book shows that the greens are not at all concerned about the environment.

They support wind farms that slice and dice wildlife, initiate bush fires, that require large areas of clear-felled land with the resultant habitat destruction and increase human emissions of carbon dioxide.

They support solar farms that do not provide clean energy, release a huge amount of toxins during manufacture, that require clear felling of large areas of land with habitat destruction and that increase human emissions of carbon dioxide.

They support biofuels that raise food costs, that increase deforestation and habitat destruction and that release large amounts of carbon dioxide into the atmosphere.

Green climate warmists think that they have a monopoly on compassion and the direction of future generations. However green groups, although registered charities, are not involved in real charitable activities. Green ideologues have hijacked public policy with junk science, anti-capitalism and self-loathing misanthropic hair-shirt propaganda and dropped these policies onto a credulous public. The results have been disastrous. The once credulous public now no longer listens and is aware that such policies have hounded blameless people out of honest jobs.

The public is aware that enquiries, commissions and other grand sounding theatrical events are stacked with nakedly political "scientific" activists with no democratic accountability, who have no contact with the punter's world and who have driven up energy, food and living costs for everyone. The public does not want to hear mendacious and deceitful spin by bullies just as they want to be able to decide what sort of light globe they purchase.

Just as an example, despite the money and media time being allocated to assorted Climate Commissions and Authorities, the level of public support for alarmist scenarios has decreased markedly in the past couple of years. Money well spent? Better to have taken it to the racetrack and punted on long-priced outsiders. Occasionally one of them actually wins.

The green's "renewable" energy policies in the Western world have resulted in fuel poverty, deaths of vulnerable people, unemployment and increased costs and, in the Third World, the perpetuation of crippling poverty and unnecessary deaths, especially of women and children. If greens were concerned about their fellow man, they would be lobbying very hard for reticulated coal-fired electricity, hydro electricity and reticulated potable water to reduce Third World poverty and deaths. They don't.

They might even raise money and build the power and water systems in the Third World themselves. They don't. The greens are knowingly immoral with their failed policies and they continue to lust for power over the average person and greedily feed at the golden public trough. The threat to the environment now comes from the greens, not from productive industry.

The greens have no knowledge of science, engineering and history, which is why they can with a straight face promote technologies that failed a long time ago. Science and technology continually create surprises, some of these become useful and all demonstrate that the science is never settled on anything. The green dream of a past utopia is demonstrably wrong. A little bit of history shows this and no one wants to go back to "the good old days".

It was technology that saved the forests, not the greens. If technology had not discovered uses for coal, then forests would have continued to be clear-felled for glass and metals manufacture. Technology has made food production far more efficient resulting in an increase in forest area. However, the green dream of biofuels continues to reduce the area of forested land.

I have sat in business class next to green political leaders in those filthy carbon dioxide emitting aeroplanes made out of an aluminium-magnesium alloy and noticed their use of metals, leather and latest technology; noticed that they quaff the best wines and eat non-organic meals, often with meat; noticed that there is a chauffeured hire car waiting just for them upon landing and noticed that they have far more benefits of the modern world than their constituents. Life's pretty good being a green politician.

Yet it is the same green politicians who argue that we ruin the planet by emitting carbon dioxide by aeroplane and car travel and by metal- and energy-making industries. They claim that we must change our lives and increase our costs. It is only wealthy countries that can afford to have green politicians and be tempted by green policies.

However, green policies have created massive debt with which most Western countries now struggle. This has been the reality check for green policies. In the Third World, green policies mean nothing and survival is everything. When these same greens live in caves and have their snouts out of the public financing trough, then we might give them a hearing. However, I recommend that we do not as they have already had a pretty good run and have shown that they have nothing to offer.

The greens are only too willing to reveal alleged environmental problems and, if they deem to offer a solution, then it is not practical and highly expensive. I wait for greens to make a sacrifice and study hard such that they are in the top 5% of school leavers, study for a first degree in science or engineering at a major university and then engage in decades of hard work to be awarded a higher degree, establish a research career and discover something that is actually beneficial to the environment. They do not. They just sit on the sidelines carping about the environment and taking all the benefits that the Western world has given them.

Can someone please tell me what green technological invention has saved millions of lives and made the planet a better place? I can only think of great technological advances that the greens try to stop thereby resulting in the deaths of millions of people in Third World countries. The greens' asinine "decarbonisation" fails all basic scientific and engineering tests but they wouldn't know because they lack basic knowledge.

If greens try to promote their message by television, radio or social media then they are hypocritically using the energy systems that they want to destroy. As I show later, the greens accept money from governments, use these monies to create unemployment and then object to governments trying to create employment. Go figure.

An understanding of how the planet works would seem necessary to be a green. It certainly is necessary to be an environmentalist. Yet greens have no idea of the past and refuse to accept that the planet has a past. This is because the past puts green cacophony into perspective. Changes in sea levels and temperature in the past have been far greater and quicker than those measured at present. The atmospheric carbon dioxide content in the past was far higher than at present yet there was no unstoppable global warming, tipping points, extreme weather and extinctions from the higher atmospheric carbon dioxide content.

The six major ice ages were all initiated when there was far more carbon dioxide in the atmosphere than now yet under the green mantra, this could not have happened. The greens ignore the past and hope it will go away. It doesn't. For most of the history of time, there has been no polar or alpine ice. When there was ice, then the ice sheets waxed and waned (as they do now) and climate was not driven by atmospheric carbon dioxide. We geologists have been studying climate change for hundreds of years and now use climate change in a practical way to provide the commodities that society uses.

Furthermore, geological models are very easily tested with drill holes and it is no wonder that geologists are amused that climate activists use models to try to predict a scary future. Models of natural processes very rarely work and the past 20 years of climate models have been tested against measurements and shown to be wrong. Very wrong.

Coal burning produces carbon dioxide. There have been great advances in scrubbing out sulphurous gases and particulates from exhaust emissions. Coal brought the Western world out of poverty and created the middle class. It is now doing the same in China, India and East Asia and yet the greens are trying to stop this process. Is this because the greens feel culturally superior to those not in the Western world? Is this because the greens want to maintain poverty and a high death rate? Is this because the greens are racist? Whatever it is, it certainly demonstrates that the greens have no conscience, do not feel responsible for their actions and are obsessed with creating unemployment in Western coal-exporting countries.

The focus on carbon dioxide as being the root of all climate evils provides a window into who really are the greens. One of the first things that children learn in school science is photosynthesis. Carbon dioxide is plant food. The past shows that with more carbon dioxide in the atmosphere, ecosystems thrived. The past shows that with warmer temperatures, ecosystems (and economies) thrived. It has yet to be shown that trace additions of a trace gas to the atmosphere drive climate change. The increase in carbon dioxide has only been good for the planet. Any horticulturist could have told the greens that warmth and carbon dioxide stimulate plant growth.

There has been increased greening of Earth and this has been measured. There is no scientific or moral reason why carbon dioxide emissions should be reduced. The objection to carbon dioxide emissions by the greens is because the green movements arose from anti-industry communist groups who wanted to slow down economic growth in the Western world and hence they use carbon dioxide and the environment as the proxy to destroy Western industry.

Green movements are now the repositories for anti-democrats, failed communists, totalitarians and the unbalanced. No wonder the co-founder of Greenpeace and writer of the Foreword to this book Dr Patrick Moore left Greenpeace in disgust after it was captured by Marxist political thugs who kicked out the real environmentalists. Greenpeace objects to any advance which will help humanity, the best example is golden rice that could save millions of children in the Third World. Are the greens really interested in enhancing the quality and length of human life?

Many green "scientists" that have chosen the role of activism. They have had short term gains in terms of fame and fortune and have irreparably damaged their institutions and science. In this book, I argue that these activist "scientists" can only live off the public purse, continue to be funded by scare campaigns and are unemployable outside their institutions. If the science of human-induced global warming is settled, as activist "scientists" claim, then these activist scientists have done their job and should be retrenched!

The founder of the Gaia hypothesis, James Lovelock, stated about green environmentalism: "It's become a religion and religions don't worry too much about facts". What was an environmental movement that was genuinely concerned about the environment has now morphed into authoritarian, anti-progress, anti-democratic, anti-human, non-scientific political pressure groups that are unelected and not responsible for their actions. If you think that this may be a little over the top, then don't listen to me, listen to the greens. Natalie Bennett from the Green Party of England and Wales stated: "Everyone who rejects the science of climate change should be fired from government: elected or unelected".

The greens were once the voice of those concerned about the environment and now have all the characteristics of a fundamentalist religion. They attack those who do not suck off the public teat or challenge the green hypocrisy. Has anyone de-programmed a fundamentalist believer? Such folk arrived at such a position with irrationality and are hardly likely to leave with a healthy dose of rationality and scepticism.

These "enlightened" greens feel morally and intellectually superior for believing the "correct" things and doing the "correct" things. The thrill and comfort of being intellectually and morally superior and fed on a diet of the supposed truth is one of the ways cults operate. There is no place for thinking, uncertainty and rationality in such groups.

If greens and media fellow travellers had to abide by the same laws as company directors and officers regarding forward projections and making demonstrably untruthful statements, then their public statements of impending catastrophes would see the courts clogged with actions against them for misleading and deceptive conduct, unsubstantiated forward projections and telling lies. By using words such as "denier", "progressive", "carbon pollution", "ocean acidification" and "renewable" the greens demonstrate that they are knowingly misleading and deceptive. The investing public is protected from financial charlatans but not from greens who destroy far more wealth than financial charlatans.

A lesson from history

There has never been a debate between those of my ilk who think that there may or may not be a little bit of warming due to human emissions of carbon dioxide and those who claim that there is dangerous warming and the consequences will be catastrophic. There has only been the echo chamber of facile groupthink and objections to this book will substantiate this opinion. It has yet been shown by green activists which part of the last 300 years of warming is natural and which part is human-induced. Without this, greens are still clutching straws about human-induced global warming.

It is well known that the greens' political negotiating position is to demand everything and not budge, no matter what the costs to the community. Greens could learn a lesson from the classic dialogue between the Athenians and Meleans. The greens behave just like the Athenians did. The Meleans were independent, refused to succumb to the expansionist Athenians' empire and, after 16 years of the 27-year long

Peloponnesian War, Athens could no longer tolerate the independence of Melos despite attempts to try to coerce Melos into an alliance.

In 416 BC, the Athenians sailed into Melos Harbour with 38 ships containing 320 archers and 2,100 heavy infantry. It was a case of negotiate or else. It was a dialogue of the deaf with the Meleans presenting their case using logic and justice and the Athenians insisting on discussing their power and expediency. The Athenians regarded power and strength as virtues rather like the greens regard themselves as holding the high moral ground. The air was toxic, not surprisingly the talks failed even before negotiations had begun.

The negotiations escalated into a war, the Athenians invaded Melos and, after many months of siege during which time the Meleans were not supported by their ally and kinsmen the Lacedæmonians, the Meleans surrendered. The Athenians killed all men of military age, sold the women and children for slaves and inhabited the island with 500 colonists. Both sides lost. Melos was destroyed, the alliances that Athens had with numerous other states collapsed as Athens could no longer be trusted.

And so too with the greens. By intransigence, stonewalling, unreasonable demands and reneging on deals, the greens just try to exert political power with no planned outcome or mutual benefit with the false premise that they hold the high moral ground. Green politicians show they can't be trusted. Centre left parties are now abandoning the greens. The Melean lesson is that strength is transitory and does not translate as negotiating power.

Follow the money

Economics of climate change

Most scientists now concede that emission reductions are either unnecessary or won't make a difference to the global climate.

Notwithstanding, some economists still try to structure profitable systems purporting to reduce human carbon dioxide emissions (e.g. carbon taxes, cap-and-trade, carbon trading).

Politicians worldwide put the cart before the horse and now there are thousands of bankers and traders who are devoting their considerable skills to designing systems that have no social purpose. The bankers supporting carbon taxes, cap-and-trade and carbon trading systems don't care one *iota* about climate change. They have been presented new and unaccountable ways of making money by governments.

Few economists have estimated the costs and benefits of some future amount of warming or cooling or changes in extreme weather. Those that have (e.g. Copenhagen Consensus Centre) calculated that the costs of adapting to a hypothetical climate change are far greater than addressing the costs if such changes ever occur at some unspecified time in the future. Economists have overlooked the benefits of a slightly higher atmospheric carbon dioxide content (e.g. enhanced photosynthesis for increased agricultural output), of slight warming (e.g. night time winter warming saves energy) and of slightly increased precipitation because warm air can dissolve more water vapour than cold air hence there is increased rainfall (e.g. agriculture).

In the Stern Report, there was a simple error with a great understating of the discount rate for future benefit. The report was trashed. The latest report from the IPCC Working Group II has concluded that global warming of 2.5°C would cost the equivalent to losing 0.2 to 2% of annual income. The Stern Review found that it would cost 5 to 20%. The latest IPCC Report claims that a mean global temperature rise of 2°C would cost 10% of the world's GDP. Whatever the figures, it is clear that it is not known and has been exaggerated in the past. The IPCC's claims are in stark contrast to global, political and economic realities.

Furthermore, there is no estimate of the costs for more expensive and more unreliable "renewable" energy. These costs are generational and already are being felt with the flight of industry from the UK and

EU to other jurisdictions. The costs and benefits of fracking to produce tight oil and gas have not been considered as have the increased energy costs by pensioning off coal and nuclear power stations and replacing them with wind and solar generators. These costs are now being felt by energy poverty and unemployment in the UK and the EU.

Australia's carbon tax was a measure to make the alleged global warming go away. It would have cost $150 billion over 10 years. During that time, the tax intended to abate 5% of Australia's carbon dioxide emissions. Australia's emissions are 1.2% of global carbon dioxide emissions. Do economists really think that it is value for money to spend $150 billion in Australia to abate only 0.06% of global carbon dioxide emissions? What does the NPV look like?

Furthermore, if Australia's carbon tax succeeded, unvalidated climate models suggest it would lower global surface temperature by 0.0007°C to 0.00007°C. Why do climate "scientists" keep telling us that there will be a forthcoming catastrophe when their own unvalidated model calculations show no such thing? If anything shows the total uselessness of climate "scientists", it is their own models.

Do economists really think that spending $150 billion over 10 years to lower global surface temperature by such an amount is a prudent investment? Australia also has a $10 billion clean energy fund on the assumption that the world is warming dangerously. There has been no warming for at least 17 years and this money could be better spent to retire debt.

If the Australian carbon tax was applied internationally with the full co-operation of China, India and the US, it would cost $317 trillion to abate one sixth of a degree of warming for the current decade. This equates to $45,000 per head of global population or 59% of global GDP. Might there be better schemes to spend money on around the world? Maybe providing clean water, reticulated electricity, education and health to most of the world would be far better and would cost far less than $45,000 per head.

What we have seen are billions of dollars of wasted public monies and the establishment of "renewable" energy and trading schemes that add costs to the public and employers. Nothing has been done to help those in need and more and more people in the Western world have collapsed into energy poverty. When it comes to rackets, "renewable energy" is much more attractive than drugs, prostitution, extortion, people smuggling, arms trade and other criminal activities because it has the government stamp of approval. Not only is it legal, one can claim to be righteous. There is no need for hit men or standover thugs as the government, free of charge, does this.

Rather than throw money at a hypothetical future problem, it would have been far more prudent for governments to think about humans adapting to a changed climate. After all, we humans have been adapting to changing climates for hundreds of thousands of years and there is nothing special about the world we live in today.

Other people's money

At the November 2013 UN climate conference in Warsaw, delegates decided that spending $1 billion a day across the world on "climate finance" was not enough. The UN Secretary General claims that more money is needed to save the world from what he calls "the greatest threat facing humanity". I would have thought that spending far less for potable water and reliable electricity in the developing world would have been a more useful venture.

What was not mentioned is that the UN's IPCC still has not demonstrated a link between carbon dioxide levels and temperature, still has not explained why the world actually flourished in the past when atmospheric carbon dioxide content was higher and buries the fact that global warming stopped 17 years ago despite human emissions of carbon dioxide increasing at an accelerated rate.

Commencing in 2014, the UN's Green Climate Fund wants to divert

$100 billion per year from developed to developing countries to "take action on climate change" and, from 2015, developed countries should spend pocket money of billions of dollars to reduce carbon dioxide emissions. This will not be enforceable but it is a valiant attempt at social engineering by the unelected to change the way the world works and to transfer wealth via their own sticky fingers.

Yet another new body was established under the UN's legal framework: the Warsaw International Mechanism for the Loss and Damage Associated with Climate Change Impacts. It is the groundwork to slug developed countries for bad weather as it is obvious that bad weather arises because developed countries have a higher standard of living. Obvious, isn't it? What is also obvious is that the same frequency of extreme weather events occurred during the cooling from 1945-1977 as in the warming from 1977-1997.

Who pays the piper?

In the US, the National Science Foundation has funded a new musical on Broadway called *The Great Immensity* to the tune of $700,000. Taxpayers are forced to fund a musical rather than leading edge scientific research. This money could have been used to fund young researchers trying to establish careers. The website of *The Great Immensity* informs us that the musical is a continent-hopping thriller with the heroine Phyllis chasing a friend who disappeared from a tropical island while on an assignment for a nature show.

The musical draws on research and interviews at Barro Colorado Island (Panama) and at Churchill (Canada) in the Arctic. It appears that the characters in the play want to share the latest environmental art, science and action to show how people around the world are having a positive impact on the big issues that we are all facing. If you think that this is thrilling, then I'll not stop you racing out to buy a ticket now. I won't see you there.

Who's who in the zoo

The Queensland and NSW governments are reviewing their Environmental Defenders' Offices following their role in a Greenpeace-led campaign to disrupt Australia's coal industry. A Greenpeace document reads: "Legal challenges can stop projects outright, or can delay them to buy time to build a much stronger movement and powerful public campaign". Governments try to create employment, especially in rural and remote areas, and some of our taxes fund Greenpeace. However, the funding used by Greenpeace is used to reduce employment in these areas and reduce government taxation and royalty revenues. Both the Queensland and NSW Environmental Defenders' Offices participated in an anti-coal campaign that aimed to block the expansion of Australia's coal export industry and, as a consequence, destroy productive jobs.

The Environmental Defenders' Offices receive Federal and State government funding, have tax deductibility status and are part of a large number of community legal centres. Grants from the Federal Attorney General's Department over 2009-2012 included $530,375 (Queensland Environmental Defenders' Office) and $526,290 (NSW Environmental Defenders' Office). There are 582 Australian environmental groups with tax deductibility status and many receive government funding. The Victorian Environmental Defenders' Office has rebadged itself to Environmental Justice Australia and will seek public donations to undertake litigation that will create unemployment.

Green groups try to pull the emotional environmental heart strings to raise funds from the credulous public and governments. Of course it sounds wonderful that green groups promote nature, want to save the wilderness and endangered species and marvel at untamed waterways. This is deceptive. If green groups argued for their real objectives, that is to destroy employment-generating rural and remote industries (especially mining, fishing, farming and forestry), discourage investment, expand government spending and controls, denigrate profit and condemn the public to the life of a noble savage then there would be very little money for green groups.

The green groups are charitable institutions that derive some income from governments which means that the taxpayer is paying for green organisations to create unemployment. It is one of life's mysteries that green political parties and their handmaidens on the left are supported by public funds yet these green groups are openly advocating job destruction. Many of the green groups below are not environmental groups and campaign for a whole basket of left political fads.

The Australian Conservation Foundation (ACF) is a charitable organisation with tax concessions. It was formed by conservatives and, in the 1980s, was captured by radical greens. The ACF campaigns on climate change and energy, "sustainable" living, northern Australia, nuclear, new economics, oceans, Murray-Darling catchment, forests and biodiversity. In 2011, they had 40,000 members with 70 staff in 5 office locations. They have affiliations with the Australian Council of Trade Unions (ACTU), World Wide Fund for Nature Australia (WWF), GetUp!, Greenpeace and the Climate Institute and historical links with the left of politics. The Australian Conservation Foundation raised $12.6 million in 2012 from government, donations, bequests, subscriptions, foundations and related parties (such as the Australian Council of Trade Unions [ACTU], WWF, Greenpeace, Environment Defenders' Office, Climate Institute) and had a small operating surplus. Decades ago I was a subscriber until I realised the unscientific misinformation and resultant economic damage and unemployment coming from the ACF.

Australian Youth Climate Coalition (AYCC) is a charitable institution with tax concessions. Their brief is to "create a world in which climate change does not threaten the future of young Australians". Maybe they should be concerned about green activists creating youth unemployment. They occupy the same office as the Climate Institute. The AYCC are active in political campaigns using members for election campaigning. They claim to have 70,000 online members, 10,000 Facebook members, funding is from governments, foundations, universities and subscriptions and they have links with the ACTU, GetUp!, Greenpeace, WWF and various left student groups. A recent email I received was from the AYCC looking for paying climate tourists to go to the Mt Everest base camp on a "Climb-it for Climate" jaunt. Your taxes at work supporting AYCC hedonism.

Friends of the Earth (FoE) is another charitable institution with tax exemptions. They are a radical green ecology group who campaign on climate justice, nanotechnology, nuclear, forests, mining, aboriginal, refugee, poverty and agriculture issues. It is ironical that their very actions create poverty. They are a national federation composed of independent member groups linked to an army of protest, green and environmental groups. In 2012, FoE declared $817,000 revenue for administration, other revenue was not disclosed. Revenue came from foreign green groups, student groups, foundations, donations, state and local governments and like-minded green groups. In 2012, their operating surplus was $201,000.

Get Up! is an Australian public company with limited tax exemptions. They are run by paid staff, interns and volunteers and campaign on gay marriage, women's rights, refugees, civil liberties, voting, the Great Barrier Reef, the Resource Tax, the Kimberleys, paper mills, live exports and banking. They are very active during elections and claim to have over 600,000 "members". These "members" are those who have signed online petitions and these "members" have no rights whatsoever regarding governance of the organisation. They are formally linked to the AYCC and have strong links with the unions and left political parties. They have run legal cases to benefit the Labor Party. Their revenue of $5 million in 2011 was from individuals, unions, student groups and family companies. They received $1.12 million from Australia's major militant union, the CFMEU, to run negative advertisements against conservative political parties.

Greenpeace (Greenpeace Australia Ltd, an affiliate of Greenpeace International) is a registered charity that has various tax exemptions.

Their specific campaign issues are on climate change, forests, GM food, oceans, whaling and nuclear. They are not afraid to break the law and are seldom held to account. Greenpeace is a limited liability company with a board of directors and about 70,000 financial supporters. In the 1990s there were about 120,000 supporters. Some 50 volunteers are voting members who elect the board. Many are connected with left political parties. Despite raising $17.1 million in 2011 and having 90 staff, Greenpeace ran at a deficit, had to be bailed out by Greenpeace International and had to shed staff.

The Climate Institute (Australia) Ltd (CIL) is a registered charity with various tax exemptions. The CIL is allegedly a "think tank", in reality it is a left political pressure group with a focus on climate change. They are affiliated with wind farm companies (some of which have strong financial links to unions) and the normal suspects such as Greenpeace, WWF, GetUp!, ACTU, ACF and AYCC. Revenue derives from foundations, government and business. In 2012, $2.2 million was raised and there was a slight deficit.

The Wilderness Society Inc. (WSI) is a charitable institution with various tax exemptions. Although allegedly a not-for-profit group, they are a relatively financially opaque federation of related organisations. They raised $13.2 million in 2012 and operated with a deficit of $75,000 (2012) with most funding from donations, subscriptions, government grants and sales. Their campaigns have been traditionally on forest issues but have now branched out into coal seam gas (which is, in effect, a product of ancient forests). The 45,000 members fund 90 staff, the WSI has intimate links with the Greens Party.

The World Wide Fund for Nature Australia (WWF) is a charitable institution with tax concessions. WWF campaigns on climate change, biodiversity and forests and has links with scientists, regulators and bureaucrats within government and business. WWF has 80 members and 40,000 supporters and also has links with some pretty radical green groups. Supporters have no input into the direction of the WWF. They have a small operating surplus from revenue of $24.2 million (2012) with funding from supporters, corporate donations, governments, bequests and foreign WWF entities.

The Environmental Defender's Offices and various green groups above have a close association with militant green activist groups. The approval process for any new mine is a long tortuous expensive process where all stakeholders can have their say. As a result of comments by stakeholders, the planned operation is normally changed or might even be abandoned.

Once approval has been granted by a formal public consultation process, various militant green groups still object, break the law as they protest against progress and job creation, disregard the safety of site workers and waste police and emergency services time. Law breaking has replaced logic and argument. These are not local groups standing up for local interests such as job creation in rural and remote areas. These militant green activists will not be seen watching their kids play football or netball on the local sporting grounds or engaged in a working bee at the local school. These militant green activists drift in from elsewhere to protest against our democratic system and job creation in the forestry, mining, oil, gas and farming industries. Who pays them?

There are at last count 582 green groups registered as charities. Where are the hospitals, refuges, kindergardens, counselling services and safe houses funded and run by these green charities? Where are the soup kitchens? Are any benevolent societies funded and run by greens? Are any green activists patrons of major charities such as the Red Cross, Lifeline, Anglicare, Brotherhood of St Laurence or Salvation Army? Do GetUp! and Greenpeace do any grass roots charitable work to help their fellow humans? Charities have traditionally tried to help people get work. Green activist groups with charitable status try to put people out of work leaving the real charities the job of picking up the pieces.

The challenge for the greens

A hypothetical question. What if the greens really wanted to do something useful for the environment? We can all think of many actions. Why not eradicate every feral animal and plant using the tried and proven methods of guns, traps, poisoning, genetic engineering, organism specific viruses and pesticides? In Australia, feral cats kill rare native birds and many rare mammals, feral goats eat everything they can, feral dogs form marauding killer packs, feral horses destroy alpine upland areas in national parks, feral foxes and rats plunder wildlife, cane toads wipe out native life, feral pigs pollute and destroy waterways, feral camels kill vegetation in sensitive desert areas, introduced fish take over from native fish and introduced plants destroy ecosystems. It is a very long list. Why don't the greens show the world that they are really concerned about the environment and actually do something? Move away from the city pulpit, get outdoors into nature and do something to make the world a better place.

Maybe the greens could spend decades obtaining knowledge and then undertake research on new sources of energy. If these ever become safe and economically competitive, then they would be ready for deployment. This would be a great successful investment by greens. Then the greens would have something to crow about and, rather than destroying jobs, the greens could actually create real jobs.

The greens object to every technological solution that would increase efficiency and lower costs, make the world a better place and bring people out of poverty yet they have no skin in the game, have not made substantial personal investments and do not lead by example. Where are the green investments in the Third World that bring potable water and reticulated electricity to the people? The greens have a large treasure chest that can easily be used to solve simple problems in the Third World.

A large number of city-based greens are high-income earners who don't put their money where their mouth is. Green fund raising is for salaries for city-based staffers, travel to huffing and puffing conferences in exotic locations and creating protest banners but not for helping people. No wonder many regard the greens as anti-human.

I wait the day when I can consult the green oracles living "sustainably" by applying their undergraduate ideology in a pristine environment using none of the trappings of the modern world. Greens in Australia could go and live in the Gibson Desert or far southern Tasmania. City-based greens would last about 2 days in the desert and, in southern Tasmania, they would freeze, get pneumonia and struggle for food. However, that is the "sustainable" life they want us to live so they should lead by example.

Those in the UK could live on the windy island of Foula in the Shetland Islands of Scotland and watch the peat form. They couldn't hug trees as there are none and they certainly could not watch the magnificent 1937 film about Foula (*Edge of the World*) as this would require electricity from fossil fuels. If they get cold or want a warm meal, then it is against their ideology to use peat, a fossil fuel, so they'll have to do something the greens have never done: invent something useful. If we had to rely on the greens, humanity would die off pretty quickly.

Those in Europe could live above the snow line in the Alps and compete with endemic fauna for moss as the principal food source. Those in North America could live in far northern Canada or Alaska and also watch the peat form and the tundra leak methane, another fossil fuel. In their spare time, they could count the increasing number of polar bears. In summer they would do battle with insects, a good source of protein. It seems that the green activist advocates preach don't do as I do, do as I say. Until the day comes when the greens live in the wilderness with no trappings of the modern world, then they are hypocrites.

Most of us take for granted that with a flick of a switch, we have energy that was produced elsewhere. This empowers us. If greens in their "sustainable" pristine locations want electricity, they can easily generate it themselves by peddling their bicycles. A generator driven by a bicycle works at about 100 watts per hour, assuming that one is physically fit and fed and watered. After ten hours of constant peddling in Nirvana, one-kilowatt hour could be produced.

By flicking a switch, we get a kilowatt-hour of electricity for 10 to 20 cents (depending upon the jurisdiction). In my house this energy is used for lighting, heating, cooling, instant hot water, food preparation and storage, cleaning clothes, communications, entertainment and security and it comes from coal. Outside the house, we use coal-fired energy for our workplaces, shopping centres, schools, hospitals and all sorts of public buildings. Maybe with the greens love of a glorious past, they could use slaves to produce the same result that I get with a flick of the switch.

Green activists attack both short- and long-term energy security. We need energy security for a future world that might not be as warm, peaceful and coherent as today. We need energy security for the inevitable global cooling and resultant food shortages. The most efficient energy for the next few hundred years, based on energy density, reserves and declining oil reserves, is to have nuclear power for electricity, coal-to-liquid for vehicle transport and dams for back up hydro electricity and food security plus gas from numerous sources and shale oil for the chemicals industry and local energy generation. However, green activism is attempting to stop nuclear power, dams, coal, gas and hydrocarbons yet green activists offer no viable tried-and-proven alternative.

Dumbed down education

More than 40 years of my life saw me in seven different universities undertaking teaching, research and administration. I especially enjoyed teaching and was sorry to retire. The world certainly changed in those decades. Early in my career, students had to achieve high public examination marks at the end of school before entry into university. Universities were for hard workers and the elite. There were not many universities and not many students. Those who were granted admission to a university had a good grasp of the basics such as English, mathematics, chemistry and physics. They read books and could spell, wrote sentences and developed arguments logically. They could be pushed very hard. Part of scholarship then was to gain knowledge, remember key facts, develop critical and analytical skills, question established authority and communicate.

Staff then were scholars and not administrators as they are now, many were poor lecturers and no matter how much a poor lecturer tried to destroy a student, the students always won because they were very good. Other lecturers were inspirational. A lot of information had to be memorised and there was an annual do-or-die written essay examination. No short answer or tick the box examination questions existed then. It was a privilege to be at university. In those times, students read for a degree. A key question then was: "How do we know what we know"? Students left university with a robust degree, they did not know much but they knew how to acquire validated knowledge and to critically evaluate information. They were employable.

Political decisions to turn every village glee club into a university a few decades ago have had profound effects. The middle class ideal has been fulfilled and all young people can now be graduates. Almost anyone with a pulse can now gain entry to a university to study for a degree in something from some institution somewhere. Universities are crammed to the rafters, class sizes have increased, student and staff quality has decreased and public universities are now businesses rather than institutions of scholarship.

Very few students can write an essay or express their thoughts in writing using knowledge and logic. At least 10% of modern university students are functionally illiterate and many more innumerate. Topics previously dealt with in the second year of a degree are now dealt with in the third or fourth year, if at all. As soon as students were pushed hard, there were howls of complaints and, in my experience, most students now just do not have the background and skills to gain a robust degree.

Many lectures are now entertainment in order to keep student numbers high and hence fill the coffers. If lecturers are not inspirational today, students are lost because personal inspiration will never come from the Internet. It is now a right to be at university, not a privilege and it is hard to fail students today even if they have not completed the required basic tasks. A single written essay examination is rare and most students are continually assessed which does not require committing a large body of knowledge to memory. A large number of degrees now are useless and it is no wonder that many graduates today are unemployed or underemployed. There is an oversupply of graduates in most disciplines.

The time has come for a handful of public and private universities to have a written matriculation entrance examination and funding based on quality rather than bums on seats. A classical education, armed with knowledge and plagued by doubt, is far superior to the current politically correct education model that is now casting young people into eternal ignorance. Even more disturbing is that these graduates do not know that they have had a poor education. The taxpayer gets poor value for money with the current university system. It is ironical that taxpayers fund elite sporting academies but there seems to be an objection to funding elite intellectual institutions. It will take a generation to correct the problems that are now firmly entrenched in our schools and universities and only by a brave government.

I have seen generations of students appear at university with very strong views on the environment. Most confuse feelings and beliefs with thinking. A few very simple questions show that they do not have the basic science to underpin their opinions and that they are just uncritically repeating ideology they heard at school. They do not even have the basic skills of logic, critical analysis and science to see the flaws in their own ideology.

A recent report in the UK by the Global Warming Policy Foundation showed green politics is all pervasive in every part of the UK school curricula. Green politics are being taught in a quasi-religious fashion, marks are awarded for following and repeating the dogma and one cannot get marks for questioning or challenging the dogma. In order to pass, a thinking student needs to be deceitful.

Early in my career, my teaching added to the students' knowledge. Later in my career, much of the teaching was providing the basic core knowledge that should have been provided at school and opening minds that had been closed by ideology. Towards the end of my university career, it was a process of playing catch up with young people who had not been taught how to think and had little basic knowledge.

One of the reasons the global warming scare campaign has been able to obtain traction is because of the dumbing down of the education system. The community is lectured to and hectored by self-appointed quasi-intellectual elites on alleged human-induced climate change rather than receiving informed opinion intertwined with modest uncertainty. Many university staff now are products of the dumbed down education system and it shows. I have seen universities evolve from institutes with great diversity to being green mouthpieces.

Universities now even donate money to green political pressure social media sites. The same universities also complain that they are underfunded. Money has usurped scholarship and this is seen in hiring and firing, student numbers, research grants and key performance indicators. Soft science is published in ideologically friendly journals by staff who have not learned logic and critical thinking. These are mechanistic contributions and are very easy to produce in large numbers. Many staff and research institutes have prospered by playing the climate game whereas those rigorous disciplines producing the graduates with the necessary skills to make the world a better place are few and far between.

Universities now have more administrators than lecturers, far too much time is spent playing pass the administrative parcel and ticking boxes and, because universities have many State and Federal masters, a large amount of time is spent on bureaucratic red and green tape answering very similar requests. One might think that vice chancellors would enforce academic standards but they now view themselves as CEOs where the bottom line is the only standard.

I occasionally met an intelligent world-wise person during my time in universities. However, I have met many more in business and outback pubs.

Your stainless steel teaspoon

A simple item of everyday life shows that greens have contributed nothing for the betterment of the world. The creation of green political parties and political coalitions with the centre left has given greens entry into government and control of the spoils rather than ensuring that all citizens enjoy a growing share of expanding economies. The centre left entered into coalitions with the greens, they supped with the Devil with a short spoon and now are politically damaged. Furthermore, the spoon wasn't even made of stainless steel.

Governments have assumed that a rise in human emissions of carbon dioxide would lead to a rise in temperature. This has shown to be wrong. This assumption has underpinned national energy policies that have left Western countries exposed to energy security risks and high costs. This could have been avoided if governments had listened to a breadth of informed opinion rather than noisy greens. This could have been avoided if journalists were sceptical and held politicians to account rather than being advocates of green ideology.

All the greens have done is increased costs, pushed more people in the Western world into fuel poverty, created unemployment, ignored crushing and solvable problems of the Third World and hypocritically moralised without setting an example. The greens want to ideologically socially engineer a brave new world without intellectual, technical and financial skills and, to date, their attempts have been very costly failures with environmental disasters paid for by the average taxpayer.

As a result of science, engineering and industry, the world has become a much better place and the environment has improved. We are beneficiaries and the humble stainless steel teaspoon is a symbol of thousands of years of experimentation, risk, trade and technology. Your humble stainless steel spoon has embedded energy and would not be possible without coal.

Many of us have had a gutful of extremist unelected political groups responsible to no one yet attacking the pillars of society that are the end result of thousands of years of social evolution. Unless the greens live in caves in the forest with no trappings of the modern world, then they should be regarded as hypocrites and treated with the disdain they deserve.

The bottom line is simple: Greens put people out of work. Will you be next?

CPSIA information can be obtained
at www.ICGtesting.com
Printed in the USA
BVHW030506140220
572296BV00001B/58